Lecture Notes in Physics

Founding Editors

Wolf Beiglböck

Jürgen Ehlers

Klaus Hepp

Hans-Arwed Weidenmüller

Volume 1046

The series Lecture Notes in Physics (LNP), founded in 1969, reports new developments in physics research and teaching - quickly and informally, but with a high quality and the explicit aim to summarize and communicate current knowledge in an accessible way. Books published in this series are conceived as bridging material between advanced graduate textbooks and the forefront of research and to serve three purposes:

- to be a compact and modern up-to-date source of reference on a well-defined topic;
- to serve as an accessible introduction to the field to postgraduate students and non-specialist researchers from related areas;
- to be a source of advanced teaching material for specialized seminars, courses and schools.

Both monographs and multi-author volumes will be considered for publication. Edited volumes should however consist of a very limited number of contributions only. Proceedings will not be considered for LNP.

Volumes published in LNP are disseminated both in print and in electronic formats, the electronic archive being available at springerlink.com. The series content is indexed, abstracted and referenced by many abstracting and information services, bibliographic networks, subscription agencies, library networks, and consortia.

Proposals should be sent to a member of the Editorial Board, or directly to the responsible editor at Springer:

Dr Lisa Scalone

lisa.scalone@springernature.com

Raghunath Sahoo

Relativistic Kinematics

A Journey in Spacetime

 Springer

Raghunath Sahoo
Department of Physics
Indian Institute of Technology Indore
Indore, Madhya Pradesh, India

ISSN 0075-8450 ISSN 1616-6361 (electronic)
Lecture Notes in Physics
ISBN 978-3-032-09509-1 ISBN 978-3-032-09510-7 (eBook)
https://doi.org/10.1007/978-3-032-09510-7

This Springer imprint is published by the registered company Springer Nature Switzerland AG
The registered company address is: Gewerbestrasse 11, 6330 Cham, Switzerland

If disposing of this product, please recycle the paper.

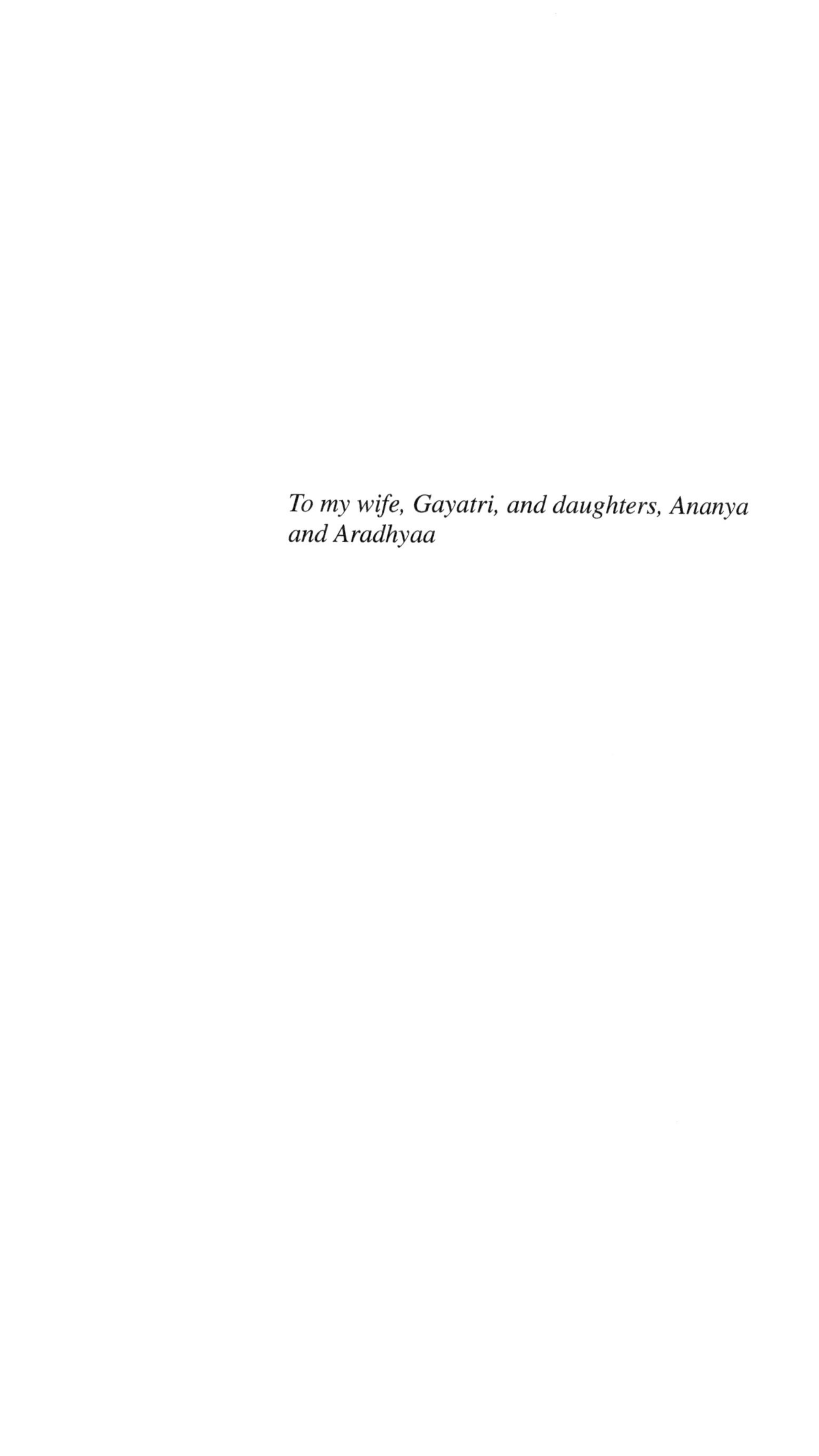

To my wife, Gayatri, and daughters, Ananya and Aradhyaa

Foreword

Albert Einstein's theory of relativity, formulated over a century ago in his 1905 *"Annus Mirabilis"*, revolutionized our understanding of space, time, and motion—serving as an essential tool in today's high-energy frontiers to understand the physics behind the (sub)atomic universe. To describe the behavior of subatomic particles and observers in the fabric of spacetime, a subject that elegantly blends geometry with physics is relativistic kinematics. This profound framework serves as a backbone of high-energy physics in the era of terra-electron-volt collisions of hadrons and nuclei to understand the creation and the evolution of the Universe.

Relativistic Kinematics: A Journey in Spacetime takes the researchers on a conceptual and mathematical voyage through this fascinating domain, properly embedded with real-world problems and pedagogy. This book is not merely a collection of equations and transformations; it is a carefully structured narrative with chosen worked-out problems that invite beginners in the field of high-energy physics to understand various underlying physical phenomena from a new and more fundamental perspective.

What makes this book a kind of its own is its clarity of exposition and its pedagogical progression. Beginning with the breakdown of classical intuition and the necessity of a new postulate-driven approach, the book advances through the core principles of Lorentz transformations, spacetime intervals, and four-vectors, ultimately exploring collisions, decays, and other dynamical processes relevant to high-energy physics in general and the quark-gluon plasma to be particular. Along the way, it offers insights into the physical significance behind the mathematics and connects formal developments with real-world phenomena, the problems a practitioner of high-energy physics encounters on his/her way to understand the dynamics of the subatomic universe.

I find this work timely and valuable—an important addition for all the young and budding researchers. It fills a critical gap between theoretical treatments and application-oriented perspectives, thereby making it an excellent resource for researchers in high-energy physics. I am aware that this book is coming after almost a decade since the author delivered this as a series of lectures in the SERC School on High-energy Physics, India-ALICE-STAR Schools, etc., and his lecture notes were widely used by everyone. I am extremely happy that, finally, a progressively refined form with a detailed structure is coming out in the form of a book, which serves as a useful resource for the future.

This is an era when the world's largest accelerator, the Large Hadron Collider, collides protons with protons at 13.6 TeV energies, and new facilities like the FAIR at GSI, Darmstadt, Germany, NICA in JINR, Dubna, Russia, and similar facilities across the globe are coming up to extensively study the phase diagram of quantum chromodyanmics (QCD). This book, I am sure, will be proved as an essential company to help and understand the new data in a spectrum of collision energies. High-energy (nuclear)physics at its edge stays highly exciting and promising with the discovered and yet to be discovered territories.

I commend and congratulate the author, Professor Raghunath Sahoo, for his lucid writing and conceptual rigor. I am confident that *Relativistic Kinematics: A Journey in Spacetime* will find its rightful place on the shelves of curious minds and scientific explorers.

Mumbai, New Delhi, India   Professor Ajit Kumar Mohanty
August 2025   Secretary, Department of Atomic Energy &
                                             Chairman, Atomic Energy Commission,
                                                              Government of India

Preface

The twentieth century witnessed a profound revolution in our understanding of space, time, and motion. Albert Einstein's theory of special and general relativity reshaped the very foundations of physics, replacing the comfortable absolutes of Newtonian mechanics with a new vision of spacetime—one that is dynamic, intertwined, and deeply connected to the geometry of the universe. Lorentz, Minkowski, and others played a vital role in our present understanding of spacetime. Within this framework, the study of relativistic kinematics stands as a cornerstone, providing the essential language through which we describe and understand the wonderworld of subatomic physics.

This book, *Relativistic Kinematics: A Journey in Spacetime*, has been written with the aim of guiding the reader through the conceptual and mathematical structure of the special theory of relativity, beginning with its classical roots and culminating in its modern applications in heavy-ion collisions. The journey starts from Galilean relativity, gradually develops the Lorentz transformations, and then extends into the four-dimensional geometry of Minkowski spacetime. Along the way, topics such as the invariance of the speed of light, time dilation, length contraction, the mathematical elegance of four-vectors, energy-momentum relations, and relativistic collisions are treated with clarity and detail, with the much-needed examples.

Beyond its theoretical framework, relativistic kinematics plays a vital role in contemporary physics. It is the foundation upon which we analyze high-energy collisions, and probe the quark-gluon plasma through extensive theoretical and experimental research. The concepts developed here are not confined to the classroom; they are the very tools used in modern particle accelerators, and the search for new physics beyond the Standard Model.

The spirit of this book is not only to present the mathematics of special relativity but also to cultivate an appreciation for its elegance in dealing with the real-world problems in high-energy heavy-ion collisions. Each chapter has been designed to build intuition through examples, diagrams, and carefully chosen problems that encourage active engagement. My hope is that students, researchers, and curious readers alike will find in these pages both a rigorous treatment of the subject and an invitation to wonder about the deeper nature of spacetime.

Writing a book is itself a journey, and this work has been enriched by countless discussions with colleagues, students, and mentors. To them I owe my deepest gratitude. My classroom lectures on special theory of relativity and lectures in various schools on relativistic kinematics of heavy-ion collisions, which heavily rely on the former, brought the idea of bringing the contents of the book to a wider use starting from undergraduate to research domains.

It is my sincere hope that this book will serve as a companion to all those embarking on their own journey through the fascinating landscape of special theory of relativity and relativistic kinematics of high-energy heavy-ion collisions.

As this is the first edition of the book, I shall appreciate any comments and suggestions to improve the content of the book, if you reach me at: Raghunath.Sahoo@cern.ch or raghunath@iiti.ac.in. The unexpected errors/typos will be posted on my webpage hosted at IIT Indore.

Indore, India   Raghunath Sahoo
August 2025

Acknowledgments I would like to put on record, Prof. Ajit Kumar Mohanty, the Chairman of the Indian Atomic Energy Commission and Secretary, Department of Atomic Energy, Government of India, for his constant encouragement and motivation to make this book possible. He was the first person, after my 2013 SERC School lecture on "Relativistic Kinematics" at the Indian Institute of Technology Madras, India, to suggest that I go for a formal write-up for the use of the research community. This encouraged me initially to go for an arXiv write-up. Later, I found that researchers whom I had never met earlier know me by name, and that is because of my arXived lecture on relativistic kinematics. This truly encouraged me to gear up for a regular book. The initial plan was to write the kinematics of heavy-ion collisions with some Monte Carlo simulations in collaboration with Prof. B.K. Nandi of IIT Bombay. We had a lot of exciting discussions together, but the project didn't see the light, possibly because of heavy engagements in teaching. I would like to thank Prof. Nandi for the initial collaborations, which, in fact, pushed the book project to some extent. Getting a dedicated time for book writing out of regular teaching time is a hard job. A source of constant support to make this project possible is my wife, Dr. Gayatri Sahu, who has been after me to bring the book project to completion. She, along with our daughters, Ananya and Aradhyaa, took regular stock of the situation and pushed constantly. The beautiful smiles of our daughters have been a constant force behind all my academic achievements, their smiles reflecting a sense of pride. No thanks can be enough to acknowledge this.

I am grateful to my teachers at Utkal University, Bhubaneswar, Prof. Niranjan Barik, Prof. L.P. Singh, and others who encouraged me to pursue a career in high-energy physics. Their teaching was at a different height, which we are yet to achieve. I would like to thank my PhD advisor, Professor D.P. Mahapatra, and my senior Professor Bedangadas Mohanty for introducing the field of high-energy heavy-ion collisions to me. At the Institute of Physics, Bhubaneswar, I was first introduced to $\eta - \phi$ of high-energy heavy-ion collisions by Prof. Mohanty. I am extremely thankful to him. His hard work and dedication in academia are a source of courage for many of us working in the field of high-energy nuclear physics. The ALICE Collaboration at CERN gave me an excellent exposure to heavy-ion physics, an outcome of which is beyond words to express—all of our collaborators carry the message of "Science Beyond Borders." Writing books and preserving the knowledge for the next generations is a different responsibility in academia. Prof. Dinesh K. Srivastava, my research collaborator, is a perfect example of such an encouraging personality— a source of inspiration for me. I shall be incomplete without mentioning late Prof. Jean Cleymans, with whom I have written close to 20 research papers. He would have been extremely happy to have a copy of this book on his table.

Teaching and research are complementary to each other and enrich the respective domains symbiotically. My teaching of nuclear and particle physics to MSc Physics students of IIT Indore, and also Experimental Methods in High Energy Physics as an elective for the PhD students, really catalyzed various domains of this book. The ALICE STAR-India collaboration gave me an excellent platform to teach the kinematics of heavy-ion collisions in its national schools organized annually. I owe a lot to my Indian ALICE-STAR collaborators.

Prof. B. Anantanarayan, IISc, Bangalore, encouraged me to write this book for Springer and brought Dr. Lisa Scalone in contact with me, who was constantly pushing the project for completion. I thank Springer Publishing for all the support. I am equally thankful to the board of editors of Springer Publishing for Physics-Books, who appreciated the book proposal by saying—the reading of the sample chapters was very pleasant, and the board liked very much the proposal for exercises within the chapters. The board thus recommended the book as a part of Springer's legendary series "Lecture Notes in Physics."

My PhD students, Mr. Suraj Prasad and Mr. Bhagyarathi Sahoo, need special mention for helping me with the excellent figures for the demonstration of the concepts and proofreading of the book. Without their help, it would not have been possible to have this piece of work on the table.

To the Lord of the Universe, Lord Jaganath, the Almighty ...

Competing Interests The author has no competing interests to declare that are relevant to the content of this manuscript.

Contents

Acronyms

STR	Special Theory of Relativity
LI	Lorentz Invariant
QCD	Quantum ChromoDynamics
LHC	Large Hadron Collider
RHIC	Relativistic Heavy Ion Collider
ALICE	A Large Ion Collider Experiment
CERN	Conseil Européen pour la Recherche Nucléaire, European Council for Nuclear Research/European Laboratory for Particle Physics
CMS	Compact Muon Solenoid
ATLAS	A Toroidal LHC ApparatuS
QGP	Quark-Gluon Plasma
LS	Laboratory System/frame
CMS/CM	Center-of-mass System/frame
LT	Lorentz Transformation
SUSY	SUper SYmmetry
CPT	Charge conjugation, Parity and Time reversal

Special Theory of Relativity

1

"When you are courting a nice girl, an hour seems like a second. When you sit on a red-hot cinder, a second seems like an hour. That's relativity."

— *Albert Einstein*

"Whether it is true or not that not more than twelve persons in all the world are able to understand Einstein's Theory, it is nevertheless a fact that there is a constant demand for information about this much-debated topic of relativity."

— *Hendrik Antoon Lorentz, The Einstein Theory of Relativity*

In this chapter, we plan to review and bring up the necessary tools used to understand the kinematics used in high-energy physics, with a special focus on heavy-ion collisions at the frontiers of nuclear physics. To do that, it becomes prudent and essential to discuss the need for high energy in dealing with the subatomic universe and also the special theory of relativity as a tool to analyse the kinematics and underlying physics.

If one considers only particles and nuclei, there are a few stable particles in nature, namely, protons, electrons, neutrinos, and photons. In terrestrial matter, there are only a limited number of nuclides available, which are in their ground states. Beyond the naturally occurring stable particles and nuclei, one needs to produce new particles and higher excited states. In addition, one needs to understand the subatomic physics, which occurs at a smaller length scale or dimension. To study the substructures of elements and composites, it is required to use probes of higher energies, which are inversely proportional to the dimension of the object to be probed. In order to understand this, let us proceed as follows.

In Particle Physics, we deal with the elementary constituents of matter. By elementary, we mean the particles have no substructure or are point-like objects. However, the elementariness depends on the spatial resolution of the probe used to

R. Sahoo, *Relativistic Kinematics*, Lecture Notes in Physics 1046,
https://doi.org/10.1007/978-3-032-09510-7_1

investigate the possible structure/sub-structure [1]. The spatial resolution is Δr if two points in an object can just be resolved as separate when they are a distance Δr apart. Assuming that the probing beams themselves consist of point-like particles like electrons or positrons, the spatial resolution is limited by the de Broglie wavelength of these beam particles, which is given by $\lambda = h/p$, where p is the beam momentum and h is Planck's constant. Hence, beams of high momentum have short de Broglie wavelengths and can have high resolution. For example, if we need to probe a dimension of 1 *Fermi* (10^{-15} meter) (let us say the inner structure of the proton, the charge radius of the proton being ~ 0.847 fm using electron beam), we need to use an electron beam of momentum 1.46 GeV, which is given by de Broglie's above expression. From optics, we know that in order to observe the structural details of linear dimension d, a wavelength *comparable* or *smaller than d* must be used:

$$\lambda \leq d. \tag{1.1}$$

This translates to a momentum requirement of

$$p \geq h/d, \tag{1.2}$$

i.e. to see smaller dimensions, probes of larger momenta or energy are required. It can be shown explicitly that if one uses an electron to probe the structure of a proton, the particle becomes relativistic– a domain where the rest mass of the particle is negligible in comparison to its energy. An explicit calculation in this example will lead to an electron momentum of 1.46 GeV/c, whereas the energy is 1.460000089 GeV. Practically, the effect of rest mass is negligible (rest mass of electron (m_0) = 0.511 MeV). Figure 1.1 shows a schematic picture where a low-energy probe fails to probe the inner structure of an object, unlike a high-energy probe with higher resolution. In addition to the above consideration, Einstein's

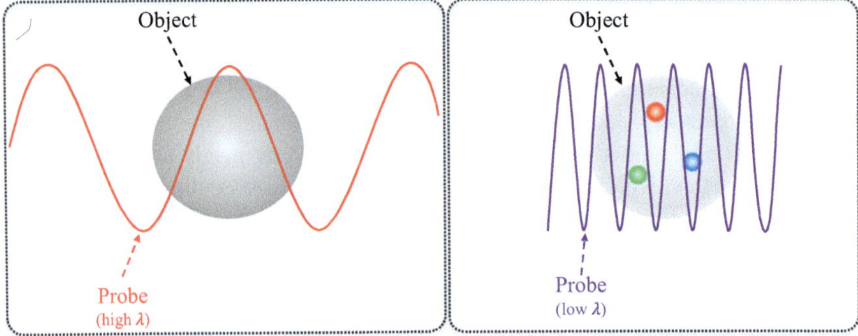

Fig. 1.1 Left: A low-energy probe (high wavelength) probing an object, Right: a high-energy probe (low wavelength) is able to probe deeper inside the object because of higher resolution

formula $E = mc^2$ helps us to produce particles of higher masses (like the massive gauge bosons, Higgs, etc.) in nature's way. Once I was asked to explain to a common man that, given the center-of-mass energy of $\sqrt{s_{NN}}$, how many pions will be produced. Without any complicated mathematics, the above formula of Einstein helps us determine the number and the answer should be: if only pion production is allowed then the number of pions produced will be $\sqrt{s_{NN}}/(m_\pi c^2)$, where m_π is the rest mass of a pion. But in real experiments, all possible particles can be produced given the available energy. In this case, for the production of exotic particles or the search for exotica like Higgs boson, SUSY particles, etc., one needs to go to much higher energies to have statistically significant numbers.

Hence, high energy is a prerequisite for:

1. The production of new particles and new states.
2. Investigation of detailed structures of the subatomic universe.

1.1 Galilean Relativity and the Need for Special Theory of Relativity

High-energy physics deals with subatomic particles, which travel nearly the speed of light. In order to describe the kinematics in high-energy physics, we need to have a level of exposure to the special theory of relativity. In order to do that with a connection to classical physics, we need to discuss some of the aspects of Galilean relativity and then migrate to the special theory while confronting the experimental reality.

Relativity connects space and time, matter and energy, electricity with magnetism. These are crucial links to understand the physical universe.

In the realm of the classical world, dealing with macroscopic objects and small velocities, more often, the consequences of the special theory of relativity is counterintuitive, like imagining length contraction and time dilation etc., which happens when a body moves at a speed closer to the speed of light (c) i.e. around 3×10^8 m/sec. Imagine a world where we live and the speed of light is reduced to 100 km/hr, and we have a clear realization of this speed, as we see a vehicle with such a speed on the highway. As a consequence, we start realizing phenomena like length contraction and time dilation in our daily events.

Let us now turn our attention to objects, speed, and the concept of relativity. Macroscopic particles in the classical world move with velocities (v) much less than the speed of light $(v/c << 1)$, whereas it is experimentally observed that subatomic particles produced in laboratories move with velocities closer to the speed of light, and Newtonian mechanics fails in such cases. While Newtonian mechanics gives no bounds to the particle velocities, in reality, it is limited by the speed of light. It is worth noting here that when we say speed of light, we refer to the speed of light in vacuum, unless otherwise it is explicitly stated. Let us consider an example to understand this better. An electron accelerated in a potential difference of 5 million volts (5 MeV energy) has a velocity equal to $0.9957c$. If we accelerate the same electron to an energy of 50 MeV, its velocity will be $0.999948c$. There is a 42% increase in the speed when there is a 10 times increase in the electron energy. However, the speed is below c. In Newtonian mechanics (kinetic energy, $K = \frac{1}{2}mv^2$), however, one would estimate the speed to be $13.9891c$, which exceeds the speed of light and hence, is unphysical. This is a clear indication that although Newtonian mechanics works well at low velocities, it fails badly for velocities closer to c. We shall show in the subsequent discussions, Einstein's 1905 theory of special relativity describes the kinematics of objects in a velocity range of $v/c \rightarrow 0$ to $v/c \rightarrow 1$, giving a generalization of Newtonian mechanics to relativistic velocities. It is interesting to mention here that Einstein in his *annus mirabilis*, 1905, came up with the landmark fundamental theories, that changed our understanding of the physical universe, namely, the Brownian motion, the photoelectric effect, and the special theory of relativity.

1.1.1 Event, Observer, and Reference Frame

Event All natural phenomena or physical occurrences take place in space and time. An event is such a phenomenon that occurs at a specific point in space and at a given instant of time. Mathematically, an event is a function of space and time.

Observer An observer is either a person or a device to record the position and time of occurrence of an event in its immediate neighbourhood. For a distant event, it has to rely on another observer.

Reference Frame A reference frame is a physical and mathematical construct to define an event by giving spatio-temporal coordinates. A reference frame in relativity usually consists of: i) an observer, ii) a coordinate system, and iii) a state of uniform (non-accelerated) motion. Or in other words, a reference frame in relativity is a system of coordinates, where an observer can assign space-time coordinates (t, x, y, z) to a physical event.

Intertial Reference Frame This frame moves with a constant (non-accelerated) velocity. The term "inertia" is derived from Newton's first law of motion– an object at rest remains at rest and an object in motion continues to move with a constant velocity (both speed and direction being constant) if no external force acts on it. As a consequence, any inertial frame of reference, which moves with a constant velocity with respect to another inertial frame (hence one can additively define a relative velocity thereof, making the complete motion a non-accelerated motion), is also an inertial frame of reference. All constant velocity motions are relative. The laws of physics, particularly Newton's first law of motion and Maxwell's equations, take the same form (invariant) in all inertial frames.

It is essential to note that the theory of relativity addresses the implications of the non-existence of a universal frame of reference. As all constant velocity motions are relative, there is no frame of reference to be assigned as *"universal"*, which is to be used everywhere. Hence, there is *"no absolute motion"*.

1.1.2   Galilean Relativity

The observational physics, in principle, started with or is attributed to the seventeenth-century Italian astronomer/physicist Galileo Galilei. Galileo was one of the first to study the mechanics of moving bodies. It is an obvious question to relate the laws of physics as observed by two observers or coordinate systems moving with a constant relative velocity with respect to each other. Galilean transformations relate the space and time coordinates of an event in one inertial frame to another moving at a constant velocity. Consider two inertial frames (as shown in Fig.1.2):

- Frame S: the stationary frame
- Frame S': moving with constant velocity $\mathbf{v}$ along the X-axis relative to S

Let an event occur at point P, with coordinates (x, y, z, t) in frame S and (x', y', z', t') in frame S'.

Assumptions

1. Time is absolute: $t' = t$
2. Space is homogeneous and isotropic.
3. Space is independent of time
4. The relative velocity between frames is constant
5. At $t = 0$, the origins of both frames coincide: $x = x' = 0$

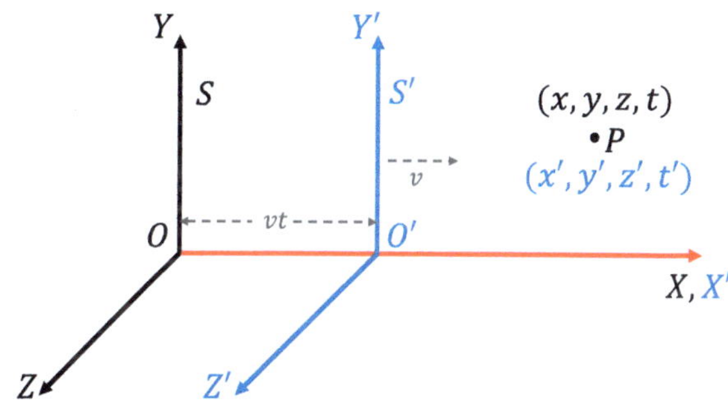

Fig. 1.2 Figure showing the frame of reference S' moving away from the reference frame S with a constant velocity v along the X−axis

Transformation of Coordinates

Since S' is moving at velocity v along the x-axis, a point at rest in S appears to move with velocity $-v$ in S'. Therefore, the position in S' is:

$$x' = x - vt$$

The motion is only along the x-axis, so the transverse coordinates remain unchanged:

$$y' = y, \quad z' = z$$

Time is the same in both frames (absolute time):

$$t' = t$$

Galilean Transformation Equations

Thus, the complete Galilean transformation from frame S to S' is:

$$\boxed{\begin{aligned} x' &= x - vt \\ y' &= y \\ z' &= z \\ t' &= t \end{aligned}} \tag{1.3}$$

To transform back from S' to S, we solve for the unprimed coordinates (or just taking v as $-v$) in order to get the inverse Galilean transformation:

$$\boxed{\begin{aligned} x &= x' + vt \\ y &= y' \\ z &= z' \\ t &= t' \end{aligned}} \tag{1.4}$$

The position vectors of the point P in both the coordinate systems are related by:

$$\boxed{\mathbf{r}' = \mathbf{r} - \mathbf{v}t} \tag{1.5}$$

1.1.2.1 Transformation of Velocity, Acceleration, and Force Under Galilean Transformation

Let the velocity of a particle as observed in frame S be $\mathbf{u} = \frac{d\mathbf{r}}{dt}$, and in frame S' be $\mathbf{u}' = \frac{d\mathbf{r}'}{dt'}$. As $t' = t$ in Galilean transformation, differentiating $\mathbf{r}' = \mathbf{r} - \mathbf{v}t$, we get:

$$\mathbf{u}' = \mathbf{u} - \mathbf{v},$$

where $\mathbf{v} = v\hat{i}$ is the relative velocity of frame S' with respect to S.

Further, to get the expression for the transformation of acceleration:
Differentiating again with respect to time:

$$\mathbf{a}' = \frac{d\mathbf{u}'}{dt'} = \frac{d\mathbf{u}}{dt} - \frac{d\mathbf{v}}{dt} = \mathbf{a},$$

since $\mathbf{v}$ is constant. Therefore,

$$\boxed{\mathbf{a}' = \mathbf{a}}.$$

This indicates that *acceleration of a particle is invariant under Galilean transformation.*

In nonrelativistic classical mechanics, the mass of a particle is invariant–independent of the frame of reference. Using Newton's second law $\mathbf{F} = m\mathbf{a}$, and since mass m is invariant:

$$\mathbf{F}' = m\mathbf{a}' = m\mathbf{a} = \mathbf{F}.$$

This means if Newton's second law of motion is valid in the frame S, it is also valid in S'. Now, if we can show that Newton's third law is also independent of the reference frame, then Newton's laws being the basis of all laws in classical mechanics, the principle of relativity would be in action. Meaning, *the laws of mechanics are the same in all inertial reference frames.*

Newton's third law of motion is a consequence of the law of conservation of momentum. Let us now proceed to show that conservation of energy and momentum are invariant under the Galilean transformation.

1.1.2.2 Conservation of Linear Momentum and Kinetic Energy Under Galilean Transformation

Conservation of Linear Momentum:

Consider a two-particle system with masses m_1 and m_2, and velocities $\mathbf{u}_1, \mathbf{u}_2$ before collision, and $\mathbf{u}'_1, \mathbf{u}'_2$ after collision in frame S.

The total momentum before and after the collision in frame S is:

$$\mathbf{p}_{\text{initial}} = m_1\mathbf{u}_1 + m_2\mathbf{u}_2, \quad \mathbf{p}_{\text{final}} = m_1\mathbf{u}'_1 + m_2\mathbf{u}'_2$$

Assuming momentum is conserved in S:

$$\mathbf{p}_{\text{initial}} = \mathbf{p}_{\text{final}}$$

Now in S', the velocities become:

$$\mathbf{u}_i^{(S')} = \mathbf{u}_i - \mathbf{v}, \quad \mathbf{u}_i'^{(S')} = \mathbf{u}_i' - \mathbf{v}$$

So, total momentum in S':

$$\mathbf{p}_{\text{initial}}^{(S')} = m_1(\mathbf{u}_1 - \mathbf{v}) + m_2(\mathbf{u}_2 - \mathbf{v}) = \mathbf{p}_{\text{initial}} - (m_1 + m_2)\mathbf{v}$$

$$\mathbf{p}_{\text{final}}^{(S')} = m_1(\mathbf{u}_1' - \mathbf{v}) + m_2(\mathbf{u}_2' - \mathbf{v}) = \mathbf{p}_{\text{final}} - (m_1 + m_2)\mathbf{v}$$

Hence, conservation of linear momentum also holds in S':

$$\mathbf{p}_{\text{initial}}^{(S')} = \mathbf{p}_{\text{final}}^{(S')}$$

Conservation of Kinetic Energy (Elastic Collision):
Kinetic energy in the frame S:

$$K_{\text{initial}} = \frac{1}{2}m_1 u_1^2 + \frac{1}{2}m_2 u_2^2, \quad K_{\text{final}} = \frac{1}{2}m_1 u_1'^2 + \frac{1}{2}m_2 u_2'^2$$

Assuming an elastic collision:

$$K_{\text{initial}} = K_{\text{final}}$$

In S', velocities are:

$$u_i^{(S')} = u_i - v \Rightarrow u_i^2 = (u_i^{(S')} + v)^2 = u_i^{(S')2} + 2vu_i^{(S')} + v^2$$

So:

$$K_{\text{initial}}^{(S')} = \frac{1}{2}m_1(u_1-v)^2 + \frac{1}{2}m_2(u_2-v)^2 = K_{\text{initial}} - v(m_1 u_1 + m_2 u_2) + \frac{1}{2}(m_1 + m_2)v^2$$

$$K_{\text{final}}^{(S')} = \frac{1}{2}m_1(u_1'-v)^2 + \frac{1}{2}m_2(u_2'-v)^2 = K_{\text{final}} - v(m_1 u_1' + m_2 u_2') + \frac{1}{2}(m_1 + m_2)v^2$$

But since momentum is conserved:

$$m_1 u_1 + m_2 u_2 = m_1 u_1' + m_2 u_2'$$

Therefore:

$$K_{\text{initial}}^{(S')} = K_{\text{final}}^{(S')}$$

Hence, both linear momentum and kinetic energy (in elastic collisions) are conserved under Galilean transformations, demonstrating their invariance in classical mechanics.

Galilean transformations are valid in classical (Newtonian) mechanics, where speeds are much smaller than the speed of light ($v \ll c$) and relativistic effects can be ignored. They fail to preserve the form of Maxwell's equations and the constancy of the speed of light, leading to the development of the Lorentz transformation in special relativity. This will be shown in the subsequent sections.

1.1.3 Galilean Transformation and the Non-invariance of the Speed of Light

Let us examine whether the speed of light remains invariant under a Galilean transformation.

Assume a light pulse is emitted from the origin at time $t = 0$. In frame S, the wavefront satisfies the spherical wave equation:

$$x^2 + y^2 + z^2 = c^2 t^2$$

This represents a sphere expanding at the speed of light c. We substitute the inverse transformation $x = x' + vt$, $y = y'$, $z = z'$, $t = t'$ into the light sphere equation:

$$(x' + vt)^2 + y'^2 + z'^2 = c^2 t'^2$$

Expanding the square:

$$x'^2 + 2x'vt + v^2 t^2 + y'^2 + z'^2 = c^2 t'^2$$

Rewriting with $t = t'$:

$$x'^2 + y'^2 + z'^2 + 2vx't' + v^2 t'^2 = c^2 t'^2$$

This is clearly **not** of the form:

$$x'^2 + y'^2 + z'^2 = c^2 t'^2$$

which would represent a light pulse expanding at speed c from the origin of frame S'.

Under the Galilean transformation, the form of the light wavefront is **not preserved**, and hence the speed of light is **not invariant**. Later, we shall discuss that; this is in contradiction to the postulate of special relativity, which states that:

The speed of light in vacuum is the same in all inertial frames of reference.

This failure of Galilean transformations led to the development of Lorentz transformations, which correctly preserve the invariance of the speed of light and the form of Maxwell's equations.

1.1.4 Non-Invariance of the Wave Equation Under Galilean Transformation

Consider the one-dimensional wave equation for a scalar field $\phi(x, t)$:

$$\frac{\partial^2 \phi}{\partial x^2} - \frac{1}{c^2} \frac{\partial^2 \phi}{\partial t^2} = 0 \tag{1.6}$$

This describes the propagation of a wave at speed c.

Under a Galilean transformation, a frame S' moves with constant velocity v relative to frame S along the x-axis, with their origins coincide at $t = t' = 0$. The transformation is:

$$x' = x - vt \tag{1.7}$$

$$t' = t \tag{1.8}$$

Thus, the inverse relations are:

$$x = x' + vt, \quad t = t'$$

Transformation of Derivatives:
We use the chain rule to relate derivatives in (x, t) to those in (x', t').
First derivatives:

$$\frac{\partial}{\partial x} = \frac{\partial x'}{\partial x} \frac{\partial}{\partial x'} + \frac{\partial t'}{\partial x} \frac{\partial}{\partial t'} = \frac{\partial}{\partial x'} \tag{1.9}$$

$$\frac{\partial}{\partial t} = \frac{\partial x'}{\partial t} \frac{\partial}{\partial x'} + \frac{\partial t'}{\partial t} \frac{\partial}{\partial t'} = -v \frac{\partial}{\partial x'} + \frac{\partial}{\partial t'} \tag{1.10}$$

Second derivatives:

$$\frac{\partial^2}{\partial x^2} = \frac{\partial^2}{\partial x'^2} \tag{1.11}$$

$$\frac{\partial^2}{\partial t^2} = v^2 \frac{\partial^2}{\partial x'^2} - 2v \frac{\partial^2}{\partial x' \partial t'} + \frac{\partial^2}{\partial t'^2} \tag{1.12}$$

Transformed Wave Equation:
Substituting into the original wave equation:

$$\frac{\partial^2 \phi}{\partial x^2} - \frac{1}{c^2} \frac{\partial^2 \phi}{\partial t^2} = \frac{\partial^2 \phi}{\partial x'^2} - \frac{1}{c^2} \left(v^2 \frac{\partial^2 \phi}{\partial x'^2} - 2v \frac{\partial^2 \phi}{\partial x' \partial t'} + \frac{\partial^2 \phi}{\partial t'^2} \right) \tag{1.13}$$

$$= \left(1 - \frac{v^2}{c^2} \right) \frac{\partial^2 \phi}{\partial x'^2} + \frac{2v}{c^2} \frac{\partial^2 \phi}{\partial x' \partial t'} - \frac{1}{c^2} \frac{\partial^2 \phi}{\partial t'^2} \tag{1.14}$$

This is clearly **not** of the same form as the original wave equation in primed coordinates:

$$\frac{\partial^2 \phi}{\partial x'^2} - \frac{1}{c^2} \frac{\partial^2 \phi}{\partial t'^2} \neq 0$$

This can easily be generalized to three dimensions to show that a wave equation is not invariant under Galilean transformation, and hence the electromagnetic wave equation. It can be explicitly shown that the laws of electromagnetism are not invariant under the Galilean transformation, which is a major drawback of Galilean relativity, limiting it to only classical mechanics— for objects moving with speed much less than the speed of light.

1.2 Electromagnetic Waves and The Speed of Light

Let us now delve into the genesis of the special theory of relativity, which will be the basis of relativistic kinematics. Some of the excitements in twentieth-century Physics are embodied in the Special Theory of Relativity (STR) and its applications. STR has its genesis from some of the paradoxes associated with electromagnetic phenomena initially studied during the nineteenth century. In 1865, James Clerk Maxwell came up with his famous set of equations concerning electric and magnetic fields, called *"Maxwell's Equations"*. Based on a completely theoretical exercise, Maxwell conjectured the existence of electromagnetic waves and that they travel at one speed– *"the speed of light, c"*. Let us now look into the idea of STR culminating from the discourse of electromagnetism. For simplicity, we adopt the SI system of units.

Based on careful experimental observations, Ampère came up with a relationship of steady current density $\mathbf{J}$ and the magnetic field $\mathbf{B}$ as

$$\nabla \times \mathbf{B} = \mu_0 \mathbf{J}, \tag{1.15}$$

where μ_0 is the permeability of free space.

A fundamental result of vector calculus says the divergence of the curl of any vector field vanishes, i.e.

$$\mathbf{\nabla}.(\mathbf{\nabla} \times \mathbf{A}) = 0, \tag{1.16}$$

for any vector $\mathbf{A}$.

Hence,

$$\mathbf{\nabla}.(\mathbf{\nabla} \times \mathbf{B}) = \mu_0(\mathbf{\nabla}.\mathbf{J}) = 0, \tag{1.17}$$

Note here that, when the left-hand side is invariably zero, the right-hand side is not always zero. Only for steady currents, $\mathbf{\nabla}.\mathbf{J} = 0$. This implies, beyond magnetostatics, Ampère's law runs into trouble. For instance, in the case of non-steady currents, *e.g.* in the process of charging a capacitor, the choice of an Ampèrian loop becomes arbitrary and hence that makes the current enclosed $I_{\text{encl}} = 0$ for a given choice of the loop in Ampère's law:

$$\oint \mathbf{B}.d\mathbf{l} = \mu_0 I_{\text{encl}}. \tag{1.18}$$

In principle, Ampère's law is derived from Biot-Savart's law, and we ought not to expect it to hold good outside magnetostatics. Maxwell fixed the issue with purely theoretical arguments.

The equation of continuity, which is in fact a mathematical description of *local charge conservation*, says:

$$\frac{\partial \rho}{\partial t} + \mathbf{\nabla}.\mathbf{J} = 0, \tag{1.19}$$

where ρ is the charge density.

Hence,

$$\mu_0(\mathbf{\nabla}.\mathbf{J}) = -\mu_0 \frac{\partial \rho}{\partial t}. \tag{1.20}$$

Gauss's law relates the charge density to the electric field:

$$\mathbf{\nabla}.\mathbf{E} = \frac{1}{\epsilon_0}\rho, \tag{1.21}$$

where ϵ_0 is the permittivity of free space.

Now,

$$-\mu_0 \frac{\partial \rho}{\partial t} = -\mu_0\epsilon_0 \frac{\partial}{\partial t}(\mathbf{\nabla}.\mathbf{E}) = -\mathbf{\nabla}.\left(\mu_0\epsilon_0 \frac{\partial \mathbf{E}}{\partial t}\right). \tag{1.22}$$

Using Eqs. (1.17), (1.19) and (1.22), one obtains:

$$\nabla \cdot (\nabla \times \mathbf{B}) = -\nabla \cdot \left(\mu_0 \epsilon_0 \frac{\partial \mathbf{E}}{\partial t} \right) \tag{1.23}$$

$$= 0 = \mu_0 \left(\nabla \cdot \mathbf{J} \right)$$

$$\Rightarrow \nabla \cdot (\nabla \times \mathbf{B}) = \nabla \cdot \left(\mu_0 \mathbf{J} + \mu_0 \epsilon_0 \frac{\partial \mathbf{E}}{\partial t} \right)$$

$$\Rightarrow \boxed{\nabla \times \mathbf{B} = \mu_0 \mathbf{J} + \mu_0 \epsilon_0 \frac{\partial \mathbf{E}}{\partial t}} \tag{1.24}$$

The extra term in Eq. (1.24) kills off the extra divergence in Ampère's law. This extra term doesn't affect magnetostatics, where $\mathbf{E}$ is constant. However, this plays a significant role in the propagation of electromagnetic waves. Like a "changing magnetic field induces an electric field (Faraday's law)", this modification of Maxwell to Ampère's law brings up: "a changing electric field induces a magnetic field". Although these prescriptions are purely theoretical, Hertz's experiment on electromagnetic waves in 1888 made the confirmation of Maxwell's theory. This extra term in Eq. (1.24) is termed as *displacement current* by Maxwell.

$$\mathbf{J}_d \equiv \epsilon_0 \frac{\partial \mathbf{E}}{\partial t}. \tag{1.25}$$

Although the term "displacement current" seems misleading, its addition of $\mathbf{J}$ in Ampère's law solves the paradox of charging a capacitor. For details, see "Introduction to Electrodynamics by D.J. Griffiths". To add to Maxwell's glory, the extra term of Maxwell in Ampère's law brought up path-breaking discoveries leading to electromagnetic waves, their propagation at the speed of light, and the birth of the special theory of relativity.

Taking the help of vector calculus, one can show that the curl of Eq. (1.24) in vacuum (regions of space where there is no charge/current, *i.e.*, $\mathbf{J} = 0$ and $\rho = 0$),

$$\nabla \times (\nabla \times \mathbf{B}) = \nabla \times \left(\mu_0 \mathbf{J} + \mu_0 \epsilon_0 \frac{\partial \mathbf{E}}{\partial t} \right)$$

$$\Rightarrow \nabla (\nabla \cdot \mathbf{B}) - \nabla^2 \mathbf{B} = \nabla \times \left(\mu_0 \epsilon_0 \frac{\partial \mathbf{E}}{\partial t} \right) = \mu_0 \epsilon_0 \frac{\partial}{\partial t} (\nabla \times \mathbf{E})$$

$$= \mu_0 \epsilon_0 \frac{\partial}{\partial t} \left(-\frac{\partial \mathbf{B}}{\partial t} \right)$$

$$\Rightarrow \boxed{\nabla^2 \mathbf{B} = \mu_0 \epsilon_0 \frac{\partial^2 \mathbf{B}}{\partial t^2}} \tag{1.26}$$

This uses Faraday's law. $\nabla \times \mathbf{E} = -\frac{\partial \mathbf{B}}{\partial t}$.

Similarly, for an electric field, one obtains

$$\nabla^2 \mathbf{E} = \mu_0 \epsilon_0 \frac{\partial^2 \mathbf{E}}{\partial t^2} \tag{1.27}$$

A general three-dimensional wave equation:

$$\nabla^2 f = \frac{1}{v^2} \frac{\partial^2 f}{\partial t^2}, \tag{1.28}$$

where $v \equiv$ velocity of the wave and f is the wave amplitude. This is also written as $\ddot{f} = v^2 \nabla^2 f$, or

$$\left(\frac{1}{v^2} \frac{\partial^2}{\partial t^2} - \nabla^2 \right) f = 0.$$

Here, the operator ∇^2 is called the Laplacian, and the bracketed operator in the above equation is termed as $\Box \equiv$ d'Alembertian.

Comparing Eqs. (1.26) and (1.27) with Eq. (1.28), one concludes that electromagnetic waves travel in vacuum with a speed:

$$v = \frac{1}{\sqrt{\mu_0 \epsilon_0}} = 3 \times 10^8 \mathrm{m/sec} = c, \tag{1.29}$$

where $c \equiv$ speed of light in vacuum.

Implication of this led to– "light is an electromagnetic wave".– an astounding revelation and path-breaking discovery in Maxwell's time.

Recall, ϵ_0 and μ_0 are the constants of Coulomb's law and Biot-Savart's law, respectively, and could be measured independently in the laboratory using electrical equipments/accessories. However, the speed of light measurement and proof of its constant value in vacuum is a completely different experiment. Surprisingly, these three independent measurements (measurements of ϵ_0, μ_0, and c) satisfy Eq. (1.29). Finally, pure theory met experiment and uncovered the beauty of nature.

Electromagnetic waves (light) always travel with the same constant speed in vacuum, irrespective of the observer's frame of reference, which led Einstein to postulate *"The speed of light is invariant"*.

1.3   Lorentz Transformation

Galilean transformation ran into a problem when it was established that $\mathbf{E}$ and $\mathbf{B}$ fields propagate as waves with the speed of light, c. Consider, in reference to Fig. 1.2, the observer O sends out a light wave towards the observer O', which is moving away at a velocity $\mathbf{v}$ along the positive $X - axis$. According to Newtonian mechanics, O' will observe the wave coming towards him with a velocity $c - v$. This will violate the principle of relativity, meaning the basic laws of physics are not invariant under Galilean transformation, as the form of Maxwell's equations will be different for the observers O and O'. For the observer O, $\mathbf{E}$ and $\mathbf{B}$ fields propagate as waves with the speed of light, c, whereas for O', it is with a velocity less than c. According to Einstein, in a fictitious scenario of an observer traveling at the speed of light would observe the speed of electromagnetic waves as zero! This is a catastrophe!

Given the Galilean transformation,

$$\mathbf{r}' = \mathbf{r} - \mathbf{v}t$$

$$t' = t, \tag{1.30}$$

if one considers the motion of O' along $X - axis$ with O and O' coinciding when $t' = t = 0$, one can show that the partial derivatives are related as:

$$\frac{\partial}{\partial x} = \frac{\partial}{\partial x'}, \quad \frac{\partial}{\partial y} = \frac{\partial}{\partial y'}, \quad \frac{\partial}{\partial z} = \frac{\partial}{\partial z'}, \quad and \quad \frac{\partial}{\partial t} = \frac{\partial}{\partial t'} - v\frac{\partial}{\partial x'}.$$

The wave equation

$$\left(\frac{1}{c^2}\frac{\partial^2}{\partial t^2} - \nabla^2 \right) E = 0 \tag{1.31}$$

under the above Galilean transformation becomes:

$$\left[\frac{1}{c^2}\left(\frac{\partial}{\partial t'} - v\frac{\partial}{\partial x'} \right)^2 - \nabla'^2 \right] E = 0. \tag{1.32}$$

This shows that Maxwell's equations are not invariant under Galilean transformations, and we need a different set of equations that will perform the required job for us. What modifications can be done to the above Galilean equations, which will ensure the invariance of the speed of light and hence the form of Maxwell's equations satisfying the relativity condition?

Referring to Fig. 1.2, we consider two inertial frames:

- $S : (x, t)$
- $S' : (x', t')$, moving at velocity v along the positive x-axis

In frame S, the wave equation for any component of the electric or magnetic field is:

$$\frac{\partial^2 \phi}{\partial x^2} + \frac{\partial^2 \phi}{\partial y^2} + \frac{\partial^2 \phi}{\partial z^2} - \frac{1}{c^2}\frac{\partial^2 \phi}{\partial t^2} = 0$$

Here, the field ϕ denotes either the **E** or **B** field of an electromagnetic wave. As the motion of the frame S' is assumed to be along the positive $X-axis$, the one dimensional wave equation in frame S is:

$$\frac{\partial^2 \phi}{\partial x^2} - \frac{1}{c^2}\frac{\partial^2 \phi}{\partial t^2} = 0$$

For the invariance of this form in the frame S', what we need is:

$$\frac{\partial^2 \phi}{\partial x'^2} - \frac{1}{c^2}\frac{\partial^2 \phi}{\partial t'^2} = 0$$

Let us take a linear transformation of the coordinates (t, x, y, z) and (t', x', y', z') of the form:

$$x' = ax + bt, \quad t' = dx + et, \quad y' = y, \quad z' = z. \tag{1.33}$$

Using the chain rule to transform the derivatives:

$$\frac{\partial}{\partial x} = a\frac{\partial}{\partial x'} + d\frac{\partial}{\partial t'}, \qquad \frac{\partial}{\partial t} = b\frac{\partial}{\partial x'} + e\frac{\partial}{\partial t'}$$

$$\frac{\partial^2}{\partial x^2} = a^2\frac{\partial^2}{\partial x'^2} + 2ad\frac{\partial^2}{\partial x'\partial t'} + d^2\frac{\partial^2}{\partial t'^2}$$

$$\frac{\partial^2}{\partial t^2} = b^2\frac{\partial^2}{\partial x'^2} + 2be\frac{\partial^2}{\partial x'\partial t'} + e^2\frac{\partial^2}{\partial t'^2}$$

Substitute into the wave equation in the unprimed frame:

$$\frac{\partial^2 \phi}{\partial x^2} - \frac{1}{c^2}\frac{\partial^2 \phi}{\partial t^2} = 0$$

$$\Rightarrow \left(a^2\frac{\partial^2}{\partial x'^2} + 2ad\frac{\partial^2}{\partial x'\partial t'} + d^2\frac{\partial^2}{\partial t'^2}\right)\phi$$

$$-\frac{1}{c^2}\left(b^2\frac{\partial^2}{\partial x'^2} + 2be\frac{\partial^2}{\partial x'\partial t'} + e^2\frac{\partial^2}{\partial t'^2}\right)\phi = 0.$$

By rearranging, we get:

$$\left(a^2 - \frac{b^2}{c^2}\right)\frac{\partial^2 \phi}{\partial x'^2} + \left(2ad - \frac{2be}{c^2}\right)\frac{\partial^2 \phi}{\partial x'\partial t'} + \left(d^2 - \frac{e^2}{c^2}\right)\frac{\partial^2 \phi}{\partial t'^2} = 0$$

The above equation to retain the same form in the frame S', the coefficients must satisfy:

$$a^2 - \frac{b^2}{c^2} = 1 \tag{1.34}$$

$$2ad - \frac{2be}{c^2} = 0 \tag{1.35}$$

$$d^2 - \frac{e^2}{c^2} = -\frac{1}{c^2} \tag{1.36}$$

Setting $x' = 0$ in the transformation equation, $x' = ax + bt$, we get $0 = avt + bt$ $\Rightarrow v = -b/a$ (refer to Fig. 1.2).

Solving the above equations for the unknown coefficients, we obtain:

$$a = \gamma, \quad b = -\gamma v, \quad d = -\frac{\gamma v}{c^2}, \quad e = \gamma, \quad \gamma = \frac{1}{\sqrt{1 - \frac{v^2}{c^2}}}$$

Now, substituting these values of the coefficients in Eq. (1.33), one obtains the Lorentz transformation equations as:

$$\boxed{\begin{aligned} x' &= \gamma(x - vt), \\ t' &= \gamma\left(t - \frac{vx}{c^2}\right), \\ y' &= y, \\ z' &= z. \end{aligned}} \tag{1.37}$$

with

$$\gamma = \frac{1}{\sqrt{1 - \frac{v^2}{c^2}}},$$

called the *"Lorentz factor"*. These Lorentz transformation equations preserve the form of the electromagnetic wave equation and ensure invariance of the speed of light, and hence the principle of relativity.

To obtain the inverse Lorentz transformation equations, one can invert these equations, expressing the unprimed coordinates in terms of the primed ones (thereby changing $+v$ to $-v$).

$$
\begin{aligned}
x &= \gamma (x' + vt'), \\
t &= \gamma \left(t' + \frac{vx'}{c^2} \right), \\
y &= y', \\
z &= z'.
\end{aligned}
\tag{1.38}
$$

This is equivalent to the frame S moving with a velocity $-v$ with respect to S'. Lorentz arrived at these equations while trying to explain the null results of the Michelson-Morley experiment, which searched for an absolute frame of reference. It was Henri Poincaré, who gave the name of Lorentz transformations to honour the Dutch physicist Hendrik A. Lorentz. As can be seen, Lorentz transformations treat space and time coordinates equally and they mix up to form a four-dimensional structure, called *"spacetime"*. In 1905, Einstein used many of the concepts, mathematical tools, and results developed by Lorentz and wrote his famous paper entitled *"On the Electrodynamics of Moving Bodies"*, which is today known as the *special theory of relativity*. Additionally, as Lorentz laid the fundamentals for the work by Einstein, this theory was originally called the *Lorentz–Einstein theory*.

Example 1.1 Show that the small velocity limit of the Lorentz transformation is the Galilean transformation

Solution 1.1 The Lorentz transformation for a boost along the x-axis is given by:

$$
\begin{aligned}
x' &= \gamma (x - vt) \\
t' &= \gamma \left(t - \frac{vx}{c^2} \right) \\
y' &= y \\
z' &= z
\end{aligned}
$$

where

$$
\gamma = \frac{1}{\sqrt{1 - \frac{v^2}{c^2}}}
$$

In the small velocity limit, $v \ll c$, one can expand the Lorenz factor γ using the binomial approximation:

$$\gamma = \left(1 - \frac{v^2}{c^2}\right)^{-1/2} \approx 1 + \frac{1}{2}\frac{v^2}{c^2} \approx 1$$

Also, $\frac{vx}{c^2} \to 0$. Substituting these into the Lorentz transformation:

$$x' \approx x - vt$$
$$t' \approx t$$
$$y' = y$$
$$z' = z$$

These are the Galilean transformation equations, hence:

> In the limit $v \ll c$, Lorentz transformations reduce to Galilean transformations.

Newtonian/Galilean relativity in classical physics is a special case of Lorentz-Einstein's special theory of relativity.

Example 1.2 Invariance of spherical light wavefront under Lorentz transformation

Solution 1.2 A light pulse emitted from the origin O, of the frame S, at $t = t' = 0$ propagates as a spherical wavefront in inertial frame S, is given by:

$$x^2 + y^2 + z^2 - c^2 t^2 = 0 \tag{1.39}$$

Let frame S' move with a velocity v along the x-axis relative to S. The Lorentz transformations between S and S' are:

$$x' = \gamma(x - vt) \tag{1.40}$$

$$t' = \gamma\left(t - \frac{vx}{c^2}\right) \tag{1.41}$$

$$y' = y \tag{1.42}$$

$$z' = z \tag{1.43}$$

where $\gamma = \dfrac{1}{\sqrt{1 - \frac{v^2}{c^2}}}$.

We use the inverse Lorentz transformations to write x and t in terms of x' and t':

$$x = \gamma(x' + vt') \tag{1.44}$$

$$t = \gamma\left(t' + \frac{vx'}{c^2}\right) \tag{1.45}$$

Substitute into Eq. (1.39):

$$
\begin{aligned}
x^2 + y^2 + z^2 - c^2 t^2 &= \gamma^2(x' + vt')^2 + y'^2 + z'^2 - c^2\gamma^2\left(t' + \frac{vx'}{c^2}\right)^2 \\
&= \gamma^2\left[x'^2 + 2vx't' + v^2 t'^2\right] + y'^2 + z'^2 \\
&\quad - c^2\gamma^2\left[t'^2 + 2\frac{vx't'}{c^2} + \frac{v^2 x'^2}{c^4}\right] \\
&= \gamma^2\left[x'^2\left(1 - \frac{v^2}{c^2}\right) + t'^2(v^2 - c^2)\right] + y'^2 + z'^2
\end{aligned}
$$

Now simplify using $\gamma^2\left(1 - \frac{v^2}{c^2}\right) = 1$:

$$x'^2 + y'^2 + z'^2 - c^2 t'^2 = 0 \tag{2}$$

Conclusion: The equation of the wave front retains the same form in S' as in S. Hence, the spherical symmetry of the wavefront is preserved under the Lorentz transformation.

Example 1.3 Invariance of the four-dimensional length element under Lorentz transformation

Solution 1.3 The spacetime interval (or Minkowski length element) is defined as:

$$ds^2 = dx^2 + dy^2 + dz^2 - c^2 dt^2 \tag{1.46}$$

Let us consider a Lorentz transformation for a boost along the x-axis:

$$x' = \gamma(x - vt) \tag{1.47}$$

$$t' = \gamma\left(t - \frac{vx}{c^2}\right) \tag{1.48}$$

$$y' = y \tag{1.49}$$

$$z' = z \tag{1.50}$$

Then, differentials transform as.

$$dx' = \gamma(dx - vdt) \tag{1.51}$$

$$dt' = \gamma\left(dt - \frac{vdx}{c^2}\right) \tag{1.52}$$

$$dy' = dy \tag{1.53}$$

$$dz' = dz \tag{1.54}$$

Now compute the transformed interval in the S' frame:

$$
\begin{aligned}
ds'^2 &= dx'^2 + dy'^2 + dz'^2 - c^2 dt'^2 \\
&= [\gamma(dx - vdt)]^2 + dy^2 + dz^2 - c^2\left[\gamma\left(dt - \frac{vdx}{c^2}\right)\right]^2 \\
&= \gamma^2(dx^2 - 2vdxdt + v^2 dt^2) + dy^2 + dz^2 \\
&\quad - c^2\gamma^2\left(dt^2 - 2\frac{v}{c^2}dxdt + \frac{v^2}{c^4}dx^2\right) \\
&= \gamma^2\left[dx^2 - 2vdxdt + v^2 dt^2 - dt^2 c^2 + 2\frac{vc^2}{c^2}dxdt - \frac{v^2 c^2}{c^4}dx^2\right] \\
&\quad + dy^2 + dz^2 \\
&= \gamma^2\left[dx^2(1 - \frac{v^2}{c^2}) + dt^2(v^2 - c^2)\right] + dy^2 + dz^2 \\
&= dx^2 + dy^2 + dz^2 - c^2 dt^2 = ds^2
\end{aligned}
$$

$$\Rightarrow \boxed{ds'^2 = ds^2}$$

Thus, the spacetime interval is invariant under Lorentz transformations. Like the
invariance of the length of a rod under rotation in spatial coordinate system, in
spacetime four-dimensional Minkowski space, the four-dimensional length element
is invariant under Lorentz transformation, which can also be shown as a rotation.

**Example 1.4 Lorentz transformation as a hyperbolic rotation in Minkowski
space**

Solution 1.4 In 2D Euclidean space, a rotation of the x-y plane around the $z-axis$
by an angle θ leaves the plane as x'-y' plane. The coordinates of the original and
rotated systems are related by:

$$x' = x\cos\theta + y\sin\theta \tag{1.55}$$

$$y' = -x \sin \theta + y \cos \theta \tag{1.56}$$

This transformation preserves the Euclidean length:

$$x'^2 + y'^2 = x^2 + y^2$$

In special relativity, spacetime is governed by the Minkowski metric. The invariant spacetime interval is:

$$s^2 = x^2 + y^2 + z^2 - c^2 t^2$$

Any transformation between inertial frames must preserve this quantity.

Taking the hyperbolic rotation ansatz, to preserve the Minkowski interval, we consider a transformation analogous to Euclidean rotation. This is a rotation of $x-axis$ and $(ict) - axis$ through an imaginary angle $\theta = i\phi$, which is in the complex $x - ict$ plane.

$$x' = x \cos(i\phi) + (ict) \sin(i\phi)$$
$$= x \cosh \phi + i(ict) \sinh \phi$$
$$= x \cosh \phi - ct \sinh \phi$$
$$y' = y$$
$$z' = z$$
$$ict' = -x \sin(i\phi) + (ict) \cos(i\phi)$$
$$= -ix \sinh \phi + ict \cosh \phi$$
$$\Rightarrow ct' = ct \cosh \phi - x \sinh \phi \tag{1.57}$$

Here, ϕ is called the *rapidity*, playing a role similar to the rotation angle. Let us now verify that the Minkowski interval is invariant under this transformation:

$$x'^2 + y'^2 + z'^2 - c^2 t'^2$$
$$= (x \cosh \phi - ct \sinh \phi)^2 + y^2 + z^2 - (ct \cosh \phi - x \sinh \phi)^2$$
$$= x^2(\cosh^2 \phi - \sinh^2 \phi) + y^2 + z^2 - c^2 t^2(\cosh^2 \phi - \sinh^2 \phi)$$
$$= x^2 + y^2 + z^2 - c^2 t^2$$

This shows that the transformation preserves the Minkowski spacetime interval. Let us now check the connection to our *"to be the most used"* Lorentz transformation equations.

We relate the rapidity ϕ to the velocity v via:

$$\tanh \phi = \frac{v}{c}$$

Define:

$$\beta = \frac{v}{c}, \quad \gamma = \frac{1}{\sqrt{1 - \beta^2}}, \quad \cosh \phi = \gamma, \quad \sinh \phi = \gamma \beta$$

Substituting these into the hyperbolic rotation equations:

$$x' = \gamma (x - vt)$$
$$t' = \gamma \left(t - \frac{vx}{c^2} \right)$$
$$y' = y$$
$$z' = z \tag{1.58}$$

These are precisely the standard Lorentz transformation equations for a boost in the x-direction. We have shown that the Lorentz transformation can be understood as a **hyperbolic rotation** in spacetime. This parallels Euclidean rotations, which preserve $x^2 + y^2 + z^2$, but in the geometry of Minkowski space. The angle of rotation becomes the *rapidity* ϕ, and trigonometric functions are replaced by their hyperbolic counterparts. In the subsequent discussions, we shall use rapidity as y, going along its experimental use.

1.4   Special Theory of Relativity

In particle physics, the particles are treated relativistically, meaning $E \approx pc \gg m_0 c^2$ and thus the special theory of relativity becomes a mathematical tool in describing particle kinematics.

- Space and time can not be treated independently as is done in Newtonian mechanics.
- Physical objects that were treated as an independent three-component vector and a scalar in non-relativistic physics mix in high-energy phenomena.
- Combined to form a four-component Lorentz vector that transforms like a space and time coordinate.

For a consistent and unified treatment, one relies on Einstein's theory of special relativity (STR), which has the following two underlying principles.

- *Invariance of velocity of light*: velocity of light in vacuum always remains as the constant c in any *inertial frame*.
- *Relativity Principle:* This requires covariance of the equations, namely, the physical law should keep its form invariant in any inertial frame of reference. In mathematical language, this amounts to the fact that physical laws have to be expressed in Lorentz tensors.

Note that the principle of relativity applies to Galilean transformation and is valid in Newtonian mechanics as well. But the invariance of the velocity of light necessitates the Lorentz transformation in changing from one inertial system to another that is moving relative to each other with constant speed.

1.5 Consequences of Lorentz Transformation

Let us now discuss some of the important consequences of the Lorentz transformation, which we shall encounter during our discussions in the subsequent chapters.

1.5.1 Lorentz-FitzGerald Length Contraction

Consider a rod at rest in frame S', moving with constant velocity v relative to frame S along the x-axis. Let the endpoints of the rod in S' be x_1' and x_2'. The proper length is:

$$L_0 = x_2' - x_1'$$

A length is called *"proper length"* when it is measured by an observer relative to whom the rod is at rest. Or in other words, *it is proper length when measured in its own frame of reference* (Fig. 1.3).

For the observed length of the rod in frame S, the observer must record the positions of both ends simultaneously, i.e., at the same time t.

Fig. 1.3 Figure showing the frame of reference S' moving away from the reference frame S with a constant velocity v along the x-axis. Length contraction of a moving rod is demonstrated

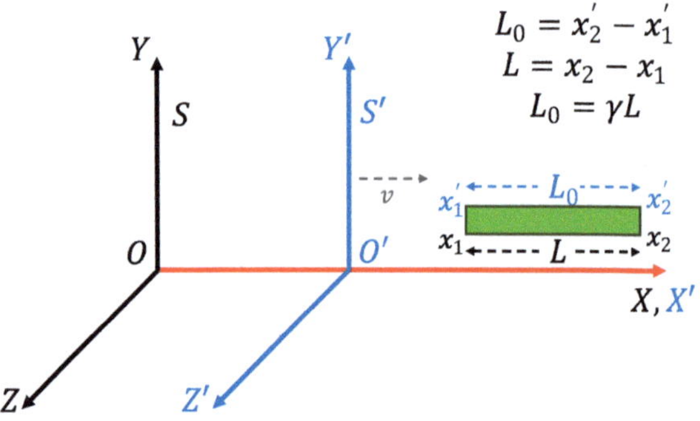

Using the Lorentz transformation:

$$x' = \gamma(x - vt)$$

We get:

$$x_1' = \gamma(x_1 - vt), \quad x_2' = \gamma(x_2 - vt)$$

Subtracting:

$$x_2' - x_1' = \gamma(x_2 - x_1) \Rightarrow L_0 = \gamma L \Rightarrow L = \frac{L_0}{\gamma}$$

$$\Rightarrow \boxed{L = L_0\sqrt{1 - \frac{v^2}{c^2}}} \tag{1.59}$$

$\sqrt{1 - \frac{v^2}{c^2}}$ is always less than unity. This indicates that $L < L_0$. The moving object appears contracted along the direction of motion. Or in other words, the length of a rod, as measured by an observer relative to whom the rod is in motion, is smaller than its proper length. This is the Lorentz–FitzGerald length contraction formula. The change is only along the boost direction, whereas the perpendicular components do not change.

One can also check this consequence of the Lorentz transformation by using the Lorentz inverse transformation when the rod is in the stationary frame S, its proper length is:

$$L_0 = x_2 - x_1$$

An observer in the S' frame measures the length of the rod through its endpoints as x_2' and x_1'. According to the inverse Lorentz transformation formula, we have:

$$x_1 = \gamma(x_1' + vt), \quad x_2 = \gamma(x_2' + vt)$$

Hence,

$$x_2 - x_1 = \gamma(x_2' - x_1') \Rightarrow L_0 = \gamma L \Rightarrow L = \frac{L_0}{\gamma}$$

The length of a rod is contracted when there is a relative motion between the rod and the observer, the proper length being the maximum length. This is also evident as the contraction depends on v^2, it is independent of the direction of the relative motion.

1.5.2   Relativity of Simultaneity

When two events occur at the same point of time at different spatial points in a given reference frame, they are called *simultaneous.* Two simultaneous events in a frame of reference are not necessarily simultaneous when seen in another frame of reference that moves at a constant velocity with respect to the first one.

Consider two events occurring at different positions x_1 and x_2 but at the same time t in the inertial frame S as shown in the Fig. 1.4. Thus:

$$t_1 = t_2 = t$$

Let t_1' and t_2' be the corresponding temporal coordinates of the two events, as measured by an observer in the frame S'.

Using the Lorentz transformation:

$$t' = \gamma \left(t - \frac{vx}{c^2} \right),$$

the time coordinates of the events in S' are:

$$t_1' = \gamma \left(t - \frac{vx_1}{c^2} \right), \quad t_2' = \gamma \left(t - \frac{vx_2}{c^2} \right),$$

This is because $t_1 = t_2 = t$, as the events are simultaneous in the frame S.

Subtracting:

$$t_2' - t_1' = \gamma \left(-\frac{v(x_2 - x_1)}{c^2} \right)$$

$$\Rightarrow \boxed{ t_2' - t_1' = -\gamma \frac{v(x_2 - x_1)}{c^2} }$$

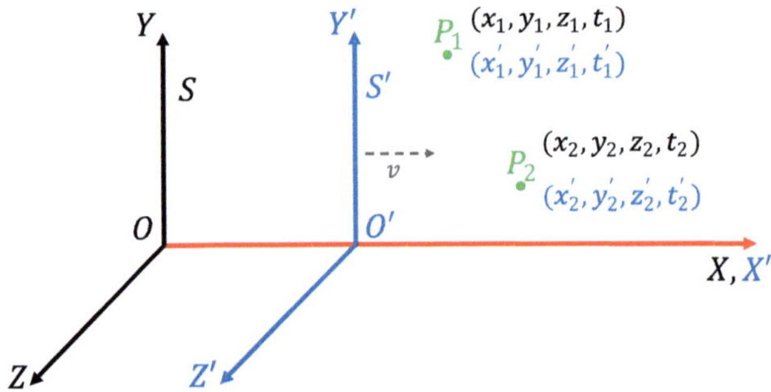

Fig. 1.4 Figure showing the frame of reference S' moving away from the reference frame S with a constant velocity v along the x-axis. Two events are shown in two inertial frames

Hence, $t_2' - t_1' \neq 0$, and this indicates that two events at two different points P_1 and P_2 which are simultaneous in the frame S, they are not found to be simultaneous for an observer in S', unless $x_1 = x_2$. This demonstrates the **relativity of simultaneity**.

1.5.3 Time Dilation

With the same configuration of the inertial frames S and S', suppose two events occur in the frame S', at times t_1' and t_2', at the same position x', as shown in Fig. 1.5. Then the time interval between the two events is:

$$\Delta t_0 \equiv \Delta t' = t_2' - t_1', \quad x_1' = x_2' = x'$$

This interval Δt_0 is called the **proper time**, as the time is measured in the clock's rest frame. Transforming to frame S, using inverse Lorentz transformation:

$$t = \gamma \left(t' + \frac{vx'}{c^2} \right),$$

we get:

$$t_1 = \gamma \left(t_1' + \frac{vx'}{c^2} \right), \quad t_2 = \gamma \left(t_2' + \frac{vx'}{c^2} \right)$$

Subtracting the above, one obtains the time interval in S

$$\Delta t = t_2 - t_1 = \gamma (t_2' - t_1') = \gamma \Delta t_0$$

$$\Rightarrow \boxed{\Delta t = \gamma \Delta t_0}$$

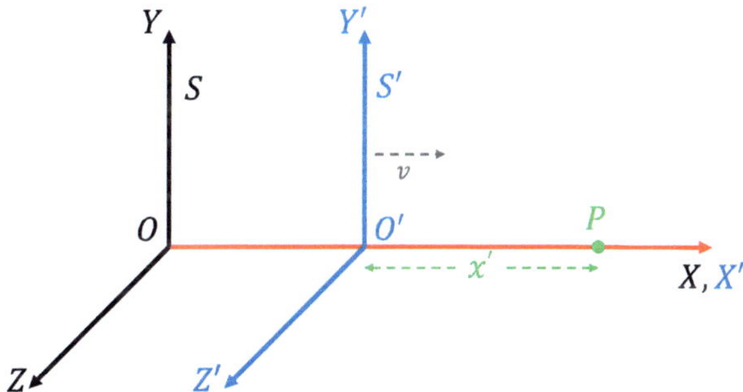

Fig. 1.5 Figure showing the frame of reference S' moving away from the reference frame S with a constant velocity v along the x-axis. Two events occur at a fixed position in S'

As $\gamma > 1$, the time between two events measured in frame S is greater than that measured in the frame S', where two events occurred at the same point (x'). This phenomenon– *to a stationary observer, the moving clock will appear to go slow*, is called **time dilation**. *The proper time is the time measured in the rest frame of the clock (where the clock is stationary) and is always the minimum.*

Example 1.5 Before setting up a muon decay experiment in your modern physics laboratory, you wanted to calculate if the cosmic ray muons, which are produced in the upper atmosphere around 8–9 km above the Earth, would reach the laboratory before they decay into an electron and two neutrinos ($\mu^- \rightarrow e^- + \bar{\nu}_e + \nu_\mu$). Given that the mean lifetime of the μ is 2.20 μs, and consider the speed as 99.8% of the speed of light.

(a) How far do the muons go before decaying? Would the muons reach the laboratory?
(b) Now consider the relativistic effect of time dilation, which would give more travel time and answer the above question.

Solution 1.5

(a) The distance travelled before the muons decay is given by:
$$d = v\tau = (0.998 \times 3 \times 10^8) \text{ m/sec} \times 2.2 \times 10^{-6} \text{ sec} = 659 \text{ m}.$$
   Note that the lifetime is the proper time (τ) of a particle, which is invariant. This calculation is done in the rest frame of the muon. In this case, the muons won't reach the laboratory. This doesn't consider relativistic time dilation.
(b) Now, let us do this in the laboratory frame. The muons travel at 99.8% of the speed of light. The Lorentz boost factor is $\gamma = \dfrac{1}{\sqrt{1-\beta^2}} = \dfrac{1}{\sqrt{1-(0.998)^2}} = 15.8$

The distance travelled as observed in the laboratory is given by:
$$d = v(\gamma\tau) = \gamma(v\tau) = 15.8 \times 659 \text{ m} = 10{,}400 \text{ m} = 10.4 \text{ km}.$$
Here we have applied the concept of "time dilation" i.e. moving clocks run slow. The time as measured in the laboratory (dt) is related to the proper time (muon lifetime), $d\tau$, by the relation $dt = \gamma d\tau$. Hence, the muons could reach the laboratory before they decay. This is one of the experimental verifications of the special theory of relativity.

Example 1.6 The Problem of Pion Decay in the Laboratory: Time Dilation and Length Contraction

A pion ($m_\pi = 273\, m_e$) decays into a muon ($m_\mu = 207\, m_e$) and a neutrino ($m_\nu = 0$) with an average lifetime of 2.6×10^{-8} sec in its own rest frame. This means half of the charged pions present at any instant of time will decay after 2.6×10^{-8} sec. Pions are produced by bombarding high-energy protons on a suitable target. If the pions leave the target with a speed close to the speed of light (say $0.99\ c$, equivalently 2.97×10^8 m/sec), at what distance from the target, the intensity of pions drop to

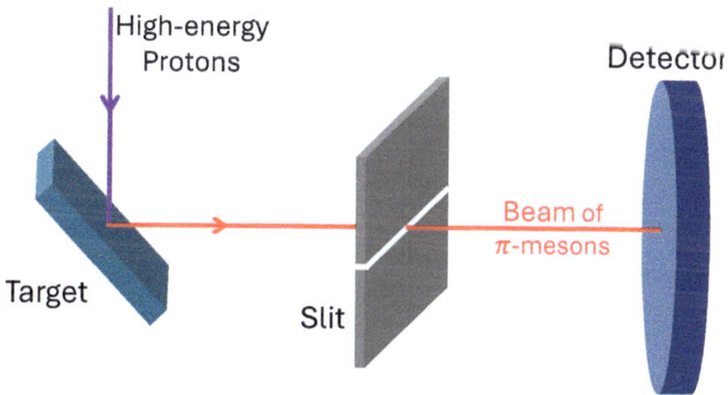

Fig. 1.6 Schematic of production and detection of pions in the laboratory

half of its original intensity? In other words, what is the suitable position to put the detectors to measure the pion intensity dropping to its half value? This is a test of relativistic physics, explained through either time dilation or length contraction (Fig. 1.6).

Solution 1.6 We have the average lifetime $\tau_0 = 2.6 \times 10^{-8}$ sec in the pion's rest frame, which is the proper time. Given the speed (v) of the pion is 2.97×10^8 m/sec, the distance travelled by a pion in its own frame is:

$$d_\pi = v \times \tau_0$$

$$= (2.97 \times 10^8)\, m/sec \,\times\, (2.6 \times 10^{-8})\, sec$$

$$= 7.7\, m$$

This, however, is not the distance at which the pion intensity is observed to be half of the original intensity. This treatment didn't consider the relativistic effect of time dilation or length contraction, as the pion velocity is close to the speed of light, and the distance between the target, the point of pion production, and the detector is measured in the laboratory frame of reference. Now, let us apply the concept of time dilation.

The average lifetime of a pion is measured by a clock in the rest frame of the pion (clock attached to the moving pion) and hence, it is the proper time of the pion, $\tau_0 = 2.6 \times 10^{-8}$ sec. Applying time dilation, the average lifetime measured by a clock in the laboratory is:

$$t = \gamma \tau_0$$

$$= \frac{2.6 \times 10^{-8}\, sec}{\sqrt{1 - \frac{(0.99c)^2}{c^2}}}$$

$$= 7.088 \times (2.6 \times 10^{-8}\, sec)$$

$$= 1.84 \times 10^{-7}\, sec$$

Hence, the distance as measured in the laboratory is:

$$d = v \times t$$
$$= (2.97 \times 10^8 \ m/sec) \times (1.84 \times 10^{-7} \ sec)$$
$$= 54.6 \ m$$

Let us now apply the concept of length contraction to counter-check the results.

We have measured the length in the laboratory (distance between the target and the detector) to be 54.6 m. The corresponding distance in the pion rest frame, which is moving with a speed 0.99 c, can be estimated by using the Lorentz-FitzGerald length contraction formula:

$$d_\pi = d/\gamma$$
$$= d\sqrt{1 - \frac{v^2}{c^2}}$$
$$= 54.6 \ m \ \times \sqrt{1 - \frac{(0.99c)^2}{c^2}}$$
$$= 54.6 \ m \ \times 0.14108$$
$$= 7.703 \ m.$$

Now, the pion average lifetime is:

$$\tau_0 = \frac{d_\pi}{v}$$
$$= \frac{7.703 \ m}{0.99c}$$
$$= \frac{7.703 \ m}{2.97 \times 10^{-8} \ m/sec}$$
$$= 2.59 \times 10^{-8} \ sec.$$

This is the observed average lifetime of pions.

1.5.4 Law of Addition of Velocities

One of the major observational setbacks to Galilean relativity was velocity of light in vacuum was not constant and it was different in different inertial frames of reference, leading to the failure of the invariance of Maxwell's equations and hence the failure of the relativity principle. Let us here derive the formula for relativistic addition of velocities and check the constancy of velocity of light under the Lorentz transformation.

As discussed, consider two inertial frames S and S', where S' is moving with velocity v along the x-axis relative to S (boost along x-axis), as shown in Fig. 1.2. Let a particle have velocity $\mathbf{u'} = (u'_x, u'_y, u'_z)$ in S', and $\mathbf{u} = (u_x, u_y, u_z)$ in S.

Here,

$\mathbf{u}$ = the velocity of a particle relative to an observer in the frame S
$\mathbf{u'}$ = the velocity of a particle relative to an observer in the frame S'
$\mathbf{v}$ = the velocity of the frame S' with respect to the frame S, along the $x - axis$.

Using the Lorentz transformation equations,

$$x' = \gamma(x - vt)$$

$$y' = y$$

$$z' = z$$

$$t' = \gamma \left(t - \frac{vx}{c^2} \right)$$

where

$$\gamma = \frac{1}{\sqrt{1 - \frac{v^2}{c^2}}}$$

Taking the differentials of the above equations:

$$dx' = \gamma(dx - vdt)$$

$$dy' = dy$$

$$dz' = dz$$

$$dt' = \gamma \left(dt - \frac{vdx}{c^2} \right)$$

So the velocity in x-direction transforms as:

$$u'_x = \frac{dx'}{dt'} = \frac{dx - vdt}{dt - \frac{vdx}{c^2}} = \frac{u_x - v}{1 - \frac{vu_x}{c^2}}$$

Similarly, the velocity transformation in y and z directions are:

$$u'_y = \frac{dy'}{dt'} = \frac{dy}{\gamma\left(dt - \frac{vdx}{c^2}\right)} = \frac{u_y}{\gamma\left(1 - \frac{vu_x}{c^2}\right)}$$

$$u'_z = \frac{dz'}{dt'} = \frac{dz}{\gamma\left(dt - \frac{vdx}{c^2}\right)} = \frac{u_z}{\gamma\left(1 - \frac{vu_x}{c^2}\right)}$$

Hence, the relativistic formula for the addition of velocities under a Lorentz boost in the x-direction is:

$$\boxed{\begin{aligned} u'_x &= \frac{u_x - v}{1 - \frac{vu_x}{c^2}} \\[2mm] u'_y &= \frac{u_y}{\gamma\left(1 - \frac{vu_x}{c^2}\right)} \\[2mm] u'_z &= \frac{u_z}{\gamma\left(1 - \frac{vu_x}{c^2}\right)} \end{aligned}} \tag{1.60}$$

While handling some kinematic problems, the inverse Lorentz transformation equations are also useful, which can be obtained by substituting $u_x \to u'_x, u_y \to u'_y, u_z \to u'_z$, and $v \to -v$:

$$\boxed{\begin{aligned} u_x &= \frac{u'_x + v}{1 + \frac{vu'_x}{c^2}} \\[2mm] u_y &= \frac{u'_y}{\gamma\left(1 + \frac{vu'_x}{c^2}\right)} \\[2mm] u_z &= \frac{u'_z}{\gamma\left(1 + \frac{vu'_x}{c^2}\right)} \end{aligned}} \tag{1.61}$$

$$\text{with} \quad \gamma = \frac{1}{\sqrt{1 - \frac{v^2}{c^2}}}$$

As one expects, in the non-relativistic domain, when $v \ll c$, so that $v/c \ll 1$ and $\gamma = 1$, Eq. (1.60) yields:

$$u'_x = u_x - v$$

$$u'_y = u_y$$

$$u'_z = u_z$$

These are the Galilean transformation equations used in classical physics.

Example 1.7   Universality of speed of light and the upper limit on speed

Solution 1.7   To simplify the problem, let us assume that the object is moving with a speed u in the frame S along the positive $x - axis$. In this case, $u_x = u$, $u_y = u_z = 0$. Similarly, with a boost along the positive $x - axis$, using Eq. (1.60) we have in S', $u'_x = u'$, $u'_y = u'_z = 0$. Using the above inverse Lorentz transformation of velocity, we obtain:

$$u = \frac{u' + v}{1 + \frac{vu'}{c^2}}, \tag{1.62}$$

and by using the Lorentz transformation of velocity, we get:

$$u' = \frac{u - v}{1 - \frac{vu}{c^2}}, \tag{1.63}$$

If we take $u' = c$, i.e., the object being a light pulse in S', then in the frame S, we have:

$$\begin{aligned}
u &= \frac{u' + v}{1 + \frac{vu'}{c^2}} \\[2mm]
&= \frac{c + v}{1 + \frac{vc}{c^2}} \\[2mm]
&= \frac{c(c + v)}{(c + v)} \\[2mm]
&= c. \tag{1.64}
\end{aligned}$$

This tells us that if a ray of light is emitted in the frame S', which is moving with a speed c with respect to the frame S, the speed of light as measured by a stationary observer in S is also c. This is *Einstein's principle of universality of speed of light, c* and was embedded inside the Lorentz transformation equations, which came much before Einstein came up with the postulates of the special theory of relativity! This is in accordance with the famous *"Michelson-Morley experiment"* and the non-existence of a universal frame of reference, *ether*. This made a huge contribution to establishing Einstein's special theory of relativity.

Example 1.8 Addition of two subluminal velocities always results in a velocity less than c

Solution 1.8 Using Eq. (1.62) for the addition of velocities:

$$u = \frac{u' + v}{1 + \frac{u'v}{c^2}}$$

Since both u' and v are less than c, $v/c << 1$, and $u'/c << 1$.

$$\Rightarrow \left(1 - \frac{v}{c}\right) \geq 0$$

and

$$\left(1 - \frac{u'}{c}\right) \geq 0$$

$$\Rightarrow \left(1 - \frac{v}{c}\right)\left(1 - \frac{u'}{c}\right) \geq 0 \Rightarrow \left(1 + \frac{vu'}{c^2} - \frac{v}{c} - \frac{u'}{c}\right) \geq 0$$

$$\Rightarrow \left(1 + \frac{vu'}{c^2}\right) \geq \frac{1}{c}(u' + v)$$

$$\Rightarrow \left(\frac{u' + v}{1 + \frac{vu'}{c^2}}\right) \leq c$$

$$\Rightarrow u \leq c \ \text{(using Eq. (1.62))}.$$

This implies that even the combination of two subluminal velocities cannot exceed the speed of light, c. Combining the results of these two examples, one observes that the speed of light c is the upper limit of all possible velocities.

Example 1.9 Angle is frame dependent

Solution 1.9 This problem is associated with the application of relativistic addition of velocity and has to do with kinematic problems in high-energy physics, while dealing with the opening angle in a two-body decay.

Consider a particle moves in frame S with velocity components (u_x, u_y), making an angle θ with the x-axis so that:

$$\tan \theta = \frac{u_y}{u_x},$$

with $u_x = u \cos \theta$ and $u_y = u \sin \theta$. Now consider a second inertial frame S' moving with velocity v along the x-axis relative to S.

The relativistic velocity transformation equations are:

$$u'_x = \frac{u_x - v}{1 - \frac{u_x v}{c^2}}$$

$$u'_y = \frac{u_y}{\gamma \left(1 - \frac{u_x v}{c^2}\right)} \qquad \text{where } \gamma = \frac{1}{\sqrt{1 - \frac{v^2}{c^2}}}$$

Then, the angle in S' is given by:

$$\tan \theta' = \frac{u'_y}{u'_x} = \frac{\dfrac{u_y}{\gamma \left(1 - \frac{u_x v}{c^2}\right)}}{\dfrac{u_x - v}{1 - \frac{u_x v}{c^2}}} = \frac{u_y}{\gamma (u_x - v)}$$

Thus,

$$\boxed{\tan \theta' = \frac{u_y}{\gamma (u_x - v)}}$$

This shows that $\theta' \neq \theta$, hence the angle is **frame-dependent** in special relativity.

The direction of motion of a particle, as measured by different observers (the center-of-momentum frame and the laboratory frame of reference, which we shall discuss later) in relative motion, appears different. Therefore:

$$\boxed{\text{Angle is not invariant under Lorentz transformations.}}$$

1.5.5 Relativistic Doppler Effect

As shown in Fig. 1.2, consider two frames of reference, S being stationary and S' moving with velocity v along the positive $X - axis$ away from S. A source of an electromagnetic signal is at the origin of the S-frame, having frequency ν. Let two signal pulses be transmitted by the source at time $t = 0$ and $t = T$. T is the true period of the electromagnetic signal. An observer stationary in the frame S' receives the signal while moving with S' with velocity v away from the frame S. As shown

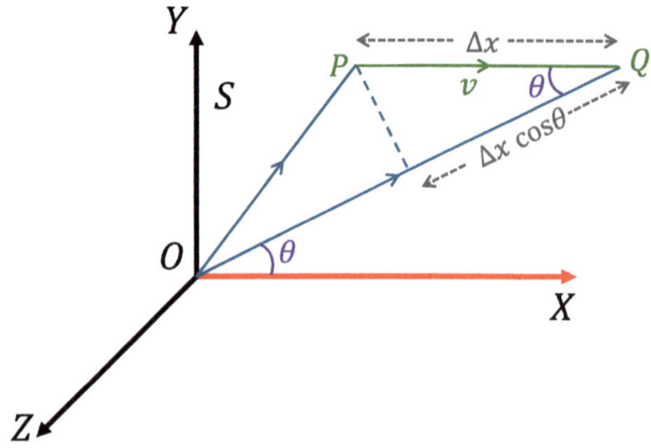

Fig. 1.7 Relativistic Doppler's effect. S'-frame is not shown here

in Fig. 1.7, let the receipt of the signals be two events P and Q at an interval of time $\Delta t'$ in S' frame. Using the Lorentz transformation, one obtains the corresponding time interval Δt in the frame S as:

$$\Delta t = \frac{\Delta t'}{\sqrt{1 - v^2/c^2}} = \gamma \Delta t'. \tag{1.65}$$

$\Delta t'$ is the proper time interval, T' for the two events, as the two events occur at a certain point in S', where the observer is at rest. As the receiver/observer is stationary in his frame (S'), $\Delta x' = 0$. From the inverse Lorentz transformations,

$$x = \gamma(x' + vt')$$

$$\Rightarrow \Delta x = \gamma(\Delta x' + v\Delta t')$$

$$\Rightarrow \Delta x = \gamma v\Delta t' \quad (\because \Delta x' = 0)$$

$$\Rightarrow \Delta x = \gamma v T'. \tag{1.66}$$

$\Delta x \cos\theta$ (as shown in the Fig. 1.7) is the distance that the second signal has to travel more than the first signal in the frame S, for which the time interval between P and Q measured in S is:

$$\Delta t = T + \frac{\Delta x \cos\theta}{c}$$

$$= T + \gamma v \Delta t' \frac{\cos\theta}{c} \quad \text{(using Eq. (1.66))} \tag{1.67}$$

Using inverse Lorentz transformations,

$$t = \gamma(t' + vx'/c^2)$$

$$\Rightarrow \Delta t = \gamma(\Delta t' + v\Delta x'/c^2)$$
$$= \gamma \Delta t' \quad (\because \Delta x' = 0)$$
$$\Rightarrow \Delta t = \gamma T'. \tag{1.68}$$

Using the above two equations:

$$\gamma T' = T + \gamma v \Delta t' \frac{\cos\theta}{c}$$
$$= T + \gamma v T' \frac{\cos\theta}{c} \quad (\because \Delta t' = T')$$
$$\Rightarrow T = \gamma T' \left(1 - \frac{v\cos\theta}{c}\right) \tag{1.69}$$

Using the time and frequency relationships, $v = 1/T$ and $v' = 1/T'$, we have:

$$\boxed{v' = \gamma v \left(1 - \frac{v\cos\theta}{c}\right)} \tag{1.70}$$

Here θ is the direction of the signals as measured in the frame S. This is the general formula for the Doppler effect, which relates the observed frequency of a signal v' in the moving frame S' to the transmitted frequency v in the frame S.

Now, if the observer moves along the positive $x - axis$ relative to the frame S, then $\theta = 0$ and $\cos\theta = 1$. Hence,

$$\boxed{v' = \gamma v \left(1 - \frac{v}{c}\right)} \tag{1.71}$$

This can be rewritten as:

$$\boxed{v' = v \left(\frac{1 - v/c}{1 + v/c}\right)^{1/2}} \tag{1.72}$$

These are the expressions for *longitudinal Doppler's effect*.

One can also write this in terms of the actual and the observed wavelengths λ and λ', using $c = v\lambda$ as:

$$\boxed{\lambda' = \lambda \left(\frac{1 + v/c}{1 - v/c}\right)^{1/2}} \tag{1.73}$$

It is clear from Eq. (1.71) that when the observer/receiver recedes away from the source emitting the electromagnetic signal, then the observed frequency is less than the original frequency. If the receiver approaches the source, the observed frequency is higher than the original frequency.

Considering only the terms with first order in v/c in Eq. (1.72), one gets:

$$\boxed{\nu' = \nu \left(1 - v/c\right)}.$$

(1.74)

This is the *non-relativistic Doppler's formula.*

Putting $\theta = 90^0$ in Eq. (1.70), we get the formula for the *transverse Doppler's effect* as:

$$\boxed{\nu' = \gamma \nu = \frac{\nu}{\sqrt{1 - v^2/c^2}}}$$

(1.75)

In the non-relativistic limit of $v/c << 1$, $\gamma \sim 1$ and the above equation gives, $\nu' = \nu$. *Unlike the special theory of relativity, there is no transverse Doppler's effect in classical physics.*

In this chapter, we have discussed the Lorentz transformation with a boost along the X-axis, along with various consequences of the special theory of relativity. For a discussion on the Lorentz transformation in an arbitrary direction, one can have a look at Appendix A. In the next chapter, we plan to discuss relativistic dynamics.

Relativistic Dynamics

2

> *"$E = mc^2$ is even better than the best poetry."*
>
> *— David Bodanis, $E = mc^2$: A Biography of the World's Most Famous Equation*

Our general intuition, driven by classical physics, says that the mass of an object is a constant quantity and doesn't depend on the velocity of the object. For example, the linear momentum of a particle of mass m_0 is defined as $m_0\mathbf{v}$. Here m_0 is independent of the particle velocity $\mathbf{v}$. On the first hand, we never use mass as m_0 while dealing with Newtonian mechanics. The fact that we have a new notation in hand, we have something else in mind to distinguish the masses as *the "rest mass" and "relativistic mass"*, which are going to do. The formulations of classical physics, when the velocity of an object is comparable to the speed of light, are not consistent with the experimental findings. When one deals with the collisions of objects moving with speeds comparable to the speed of light, one may expect that the rest mass of the object and hence the non-relativistic momentum won't be conserved, as the collisions will soon be termed as *"inelastic"*, or particle-producing collisions. Usually in Newtonian mechanics, one deals with collisions, which are called *"elastic"*, where the total rest mass is a conserved quantity. It can be shown that the non-relativistic momentum, $m_0\mathbf{v}$, is not a conserved quantity when the object moves with relativistic speeds. However, the conservation of momentum is a fundamental law of nature coming from the homogeneity of space. As a solution to this catastrophe, the definition of momentum needs to be modified to deal with the relativistic speed of the objects, and hence the law of conservation of momentum. As a consequence, we shall shortly see that, mass of an object varies with velocity and, most importantly, Einstein's famous mass-energy conversion relation, $E = mc^2$ emerges.

© The Author(s), under exclusive license to Springer Nature Switzerland AG 2025
R. Sahoo, *Relativistic Kinematics*, Lecture Notes in Physics 1046,
https://doi.org/10.1007/978-3-032-09510-7_2

2.1 Relativity of Mass

We shall use the law of conservation of momentum and the Lorentz transformation of velocity to derive the formula for the variation of mass with velocity. Here, the mass of the object m is a function of its velocity u, so that for $u = 0$, the object is at rest and its rest mass is m_0.

As shown in Fig. 2.1, consider an inelastic collision of two identical objects of mass m_1 and m_2 happening in the inertial frame S', which is receding away from another inertial frame S along the common positive $x - axis$ with a velocity v. Let u' and $-u'$ be the velocities of the objects in S'-frame along the $x' - axis$, as they move opposite to each other. The masses of the objects are equal, and hence the momenta are opposite in the S' frame. This leads to net momentum zero after the collision with mass $(m_1 + m_2)$. The objects come to complete rest in the S' frame after the collision. This is seen from the frame S as $(m_1 + m_2)$ a single object, which is moving with velocity v. m_1 and m_2 are the masses, and u_1 and u_2 are their corresponding velocities as observed from the frame S.

Now applying the law of conservation of linear momentum to the frame S, we get:

$$m_1 u_1 + m_2 u_2 = (m_1 + m_2)v. \tag{2.1}$$

According to inverse Lorentz transformation of velocity, u_1 and u_2 are given by:

$$u_1 = \frac{u' + v}{1 + \frac{u'v}{c^2}} = \frac{u' + v}{1 + x}, \tag{2.2}$$

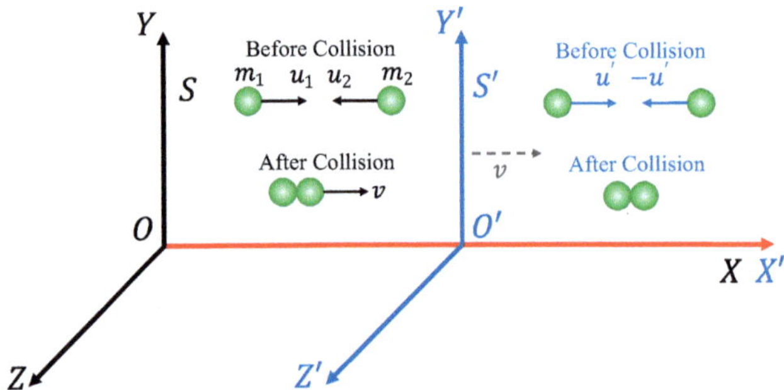

Fig. 2.1 A two body collision happening in S'-frame as observed by frame S. S' is moving away from S with a constant velocity v along the $X-$axis

and

$$u_2 = \frac{-u' + v}{1 - \frac{u'v}{c^2}} = \frac{-u' + v}{1 - x}, \tag{2.3}$$

with $x = u'v/c^2$.

Putting the values of u_1 and u_2 in Eq. (2.1):

$$m_1 \left(\frac{u' + v}{1 + x} \right) + m_2 \left(\frac{-u' + v}{1 - x} \right) = (m_1 + m_2)v$$

$$\Rightarrow m_1 \left(\frac{u' + v}{1 + x} - v \right) = m_2 \left(v - \frac{-u' + v}{1 - x} \right)$$

$$\Rightarrow m_1 \left(\frac{u' - vx}{1 + x} \right) = m_2 \left(\frac{u' - vx}{1 - x} \right)$$

$$\Rightarrow \frac{m_1}{m_2} = \frac{1 + x}{1 - x} \tag{2.4}$$

To evaluate the right side of the above equation, let us proceed as follows:

Squaring Eq. (2.2) and subtracting c^2 from both the sides:

$$u_1^2 - c^2 = \left(\frac{u' + v}{1 + x} \right)^2 - c^2$$

$$= \frac{u'^2 + v^2 + 2u'v}{(1 + x)^2} - c^2$$

$$= \frac{u'^2 + v^2 + 2u'v - c^2 - x^2c^2 - 2xc^2}{(1 + x)^2}$$

$$= \frac{u'^2 + v^2 - c^2 - x^2c^2}{(1 + x)^2} \qquad (\because x = u'v/c^2) \tag{2.5}$$

Similarly, squaring Eq. (2.3) and subtracting c^2 from both the sides:

$$u_2^2 - c^2 = \left(\frac{-u' + v}{1 - x} \right)^2 - c^2$$

$$= \frac{u'^2 + v^2 - 2u'v}{(1 - x)^2} - c^2$$

$$= \frac{u'^2 + v^2 - 2u'v - c^2 - x^2c^2 + 2xc^2}{(1 - x)^2}$$

$$= \frac{u'^2 + v^2 - c^2 - x^2c^2}{(1 - x)^2} \qquad (\because x = u'v/c^2) \tag{2.6}$$

Diving Eq. (2.6) by Eq. (2.5), we get:

$$\frac{u_2^2 - c^2}{u_1^2 - c^2} = \frac{(1+x)^2}{(1-x)^2}$$

$$\Rightarrow \frac{(1+x)}{(1-x)} = \frac{\sqrt{1 - u_2^2/c^2}}{\sqrt{1 - u_1^2/c^2}} \tag{2.7}$$

Putting this in Eq. (2.4):

$$\frac{m_1}{m_2} = \frac{\sqrt{1 - u_2^2/c^2}}{\sqrt{1 - u_1^2/c^2}}$$

$$\Rightarrow m_1\sqrt{1 - u_1^2/c^2} = m_2\sqrt{1 - u_2^2/c^2} \tag{2.8}$$

This equation to hold good for all values of u_1 and u_2, and should be equal to a constant. Now, if we assume the velocity of the second object is observed to be zero by the frame-S, i.e., $u_2 = 0$, then the mass m_2 is the rest mass of the identical objects, m_0. This constant is then identified to be the rest mass, m_0, which is also called the *"proper mass"* of the object. This is measured in the inertial reference frame where the object is at rest. Hence,

$$m_1 = \frac{m_0}{\sqrt{1 - u_1^2/c^2}}$$

and

$$m_2 = \frac{m_0}{\sqrt{1 - u_2^2/c^2}}$$

In general, the mass of a moving object, m, which is a function of velocity v, is given by

$$\boxed{m = \frac{m_0}{\sqrt{1 - v^2/c^2}} = \gamma m_0} \tag{2.9}$$

As $\gamma > 1$, $m > m_0$. The mass of a moving body is heavier than its rest mass by a factor equal to the Lorentz factor, γ. When the object moves with non-relativistic speed, i.e., $v \ll c$, then $\gamma \sim 1$ and the relativistic mass becomes equal to the rest mass— $m \simeq m_0$. On the other hand, when the speed nears the speed of light, i.e., $v \to c$, $\sqrt{1 - v^2/c^2} \to 0$ and hence, $m \to \infty$. This means the relativistic

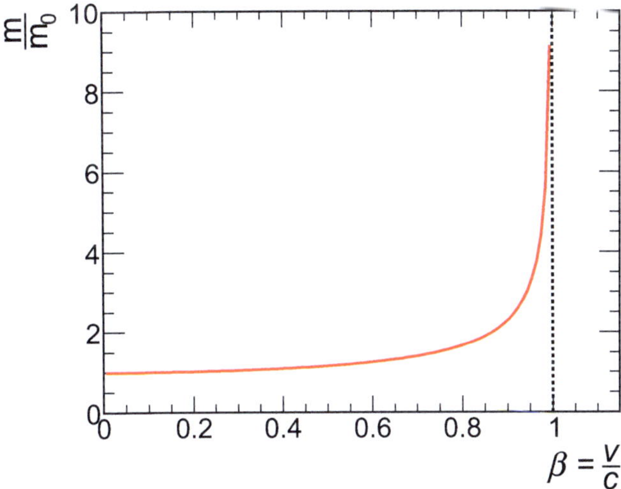

Fig. 2.2 Increase of mass with speed, which asymptotically approaches infinity(unphysical), when the speed approaches the speed of light

mass increases boundlessly and is hence unphysical. This goes well in line with the ultimate speed limit of a material object– a particle of finite mass can not travel with speed equal to the speed of light. This is one of the reasons, in particle accelerators, achieving higher speed and hence higher energy becomes increasingly difficult because of the increase in the relativistic mass of the particle. The variation of mass with speed is shown in Fig. 2.2.

When we say, no material particle can travel with or more than the speed of light, this is the speed of light in vacuum, unless otherwise, it is explicitly mentioned. It should be noted here that the case of a material medium is different, and the speed of light is smaller than its speed in a vacuum. Energetic particles like electrons in a material medium can travel faster than light in the same medium, where they emit Cherenkov radiation. This doesn't violate the theory of relativity.

The variation of mass with velocity was independently proved in 1909, four years after the conjecture of the special theory of relativity by Albert Einstein, by two physical chemists, Gilbert N. Lewis and Richard C. Tolman, at the Massachusetts Institute of Technology, USA. Lewis and Tolman used elementary arguments based on two *"Gedankenexperiment"* (thought experiments) involving light travelling between mirrors to prove the hypothesis of variation of mass with velocity, but not via the Lorentz transformation. This hypothesis was confirmed experimentally by Kaufmann, Bucherer, and Guye, and Lavanchy with high accuracy using high-speed electrons (β-rays).

Quoting Richard Feynman on the increase in mass of an object with velocity:

For those who want to learn just enough about it so that they can solve problems, that is all there to the theory of relativity – it just changes Newton's laws by introducing a correction factor to the mass.

Example 2.1   Rest mass of Photon

Solution 2.1 A photon is the quantum of light and the mediating gauge boson of electromagnetic interaction. It travels at the speed of light. The relativistic expression for momentum is $p = \gamma m_0 v$. Using the well-known de Broglie's equation and quantum theory, the momentum of a photon of wavelength λ is $p = \frac{h}{\lambda}$, where h is Planck's constant. Hence,

$$\frac{h}{\lambda} = \gamma m_0 v$$

$$\Rightarrow m_0 = \frac{h}{v\lambda}\sqrt{1 - v^2/c^2} \tag{2.10}$$

For photon, $v = c$, which implies:

$$\Rightarrow \boxed{m_0 = 0}. \tag{2.11}$$

2.2   Relativistic Momentum and Force

The linear momentum ($\mathbf{p}$) of a particle with rest mass, m_0 moving with velocity $\mathbf{v}$ is defined by:

$$\boxed{\mathbf{p} = m\mathbf{v} = \gamma m_0 \mathbf{v}.} \tag{2.12}$$

When $v \to c$, $\mathbf{p} \to \infty$: the speed of a particle approaching the speed of light, the relativistic momentum of a particle becomes infinite. Further, for the non-relativistic reduction, as expected, for non-relativistic velocities, when $v \ll c$, $\mathbf{p} = m_0\mathbf{v}$, is the classical form of linear momentum.

The law of conservation of momentum for an isolated ensemble of n-particles takes the form:

$$\mathbf{p} = \sum_{i=1}^{n} m_i \mathbf{v_i} = \sum_{i=1}^{n} \gamma_i m_{0i} \mathbf{v_i} = constant, \tag{2.13}$$

and can be shown that it is valid in the special theory of relativity. However, contrary to the case of classical mechanics, in the special theory of relativity, the force acting on a particle can not be defined as mass multiplied by the acceleration of the particle.

It should be defined as the time rate of change of momentum in line with Newton's original definition of force.

Newton's second law of motion connects the net force acting on a particle with the time rate of change of momentum:

$$\mathbf{F} = \frac{d\mathbf{p}}{dt}.$$
(2.14)

This equation is valid in the relativistic domain if $\mathbf{p}$ is the relativistic momentum of the particle.

$$\mathbf{F} = \frac{d\mathbf{p}}{dt} = \frac{d(m\mathbf{v})}{dt} = \frac{d}{dt}(\gamma m_0 \mathbf{v}).$$
(2.15)

It should be noted here that the change in the definition of mass has now modified the definition of momentum and hence, Newton's second law of motion.

Now, taking $c = 1$, which comes in the system of natural units,

$$\frac{d\gamma}{dt} = \frac{d}{dt}\left(\frac{1}{\sqrt{1 - v^2}}\right) = \frac{v\dot{v}}{(1 - v^2)^{3/2}} = \gamma^3 va,$$
(2.16)

where, $a = \frac{dv}{dt} = \dot{v}$ is the acceleration of the particle. With m_0 being constant,

$$\begin{aligned}
F &= \frac{d}{dt}(\gamma m_0 v) \\
&= m_0(\dot{\gamma} v + \gamma \dot{v}) \\
&= m_0 \gamma^3 v^2 a + m_0 \gamma \dot{v} \\
&= m_0 \gamma a(\gamma^2 v^2 + 1) \\
\Rightarrow F &= \gamma^3 m_0 a \quad (\because \gamma^2 v^2 + 1 = \gamma^2).
\end{aligned}$$

Hence,

$$\boxed{\mathbf{F} = \gamma^3 m_0 \mathbf{a}}.$$
(2.17)

Here we have considered force in one dimension for simplicity, and the force is parallel to the velocity. *This shows that, unlike classical mechanics, the force acting on a body is not proportional to acceleration.*

Example 2.2 Show that force in special theory of relativity, is in general, not proportional to acceleration

Solution 2.2 Newton's second law of motion gives:

$$
\mathbf{F} = \frac{d\mathbf{p}}{dt} = \frac{d(\gamma m_0 \mathbf{v})}{dt} = \frac{dm\mathbf{v}}{dt}
$$

$$
= m\frac{d\mathbf{v}}{dt} + \mathbf{v}\frac{dm}{dt} \tag{2.18}
$$

Einstein's mass-energy relation gives $E = mc^2$. Hence,

$$
\frac{dm}{dt} = \frac{1}{c^2}\frac{dE}{dt}
$$

$$
= \frac{1}{c^2}\frac{d(T + m_0 c^2)}{dt}
$$

$$
= \frac{1}{c^2}\frac{dT}{dt} \tag{2.19}
$$

Here, T is the kinetic energy.

Further, using the work-energy principle:

$$
\frac{dT}{dt} = \frac{d}{dt}(\mathbf{F}.d\mathbf{x})
$$

$$
= \mathbf{F}.\frac{d\mathbf{x}}{dt} = \mathbf{F}.\mathbf{v}
$$

Hence,

$$
\frac{dm}{dt} = \frac{1}{c^2}\mathbf{F}.\mathbf{v}. \tag{2.20}
$$

Using this in Eq. (2.18) we get,

$$
\mathbf{F} = m\frac{d\mathbf{v}}{dt} + \mathbf{v}\left(\frac{1}{c^2}\mathbf{F}.\mathbf{v}\right)
$$

$$
= m\mathbf{a} + \frac{\mathbf{v}}{c^2}\mathbf{F}.\mathbf{v}
$$

$$
\Rightarrow \mathbf{a} = \frac{\mathbf{F}}{m} - \left(\frac{1}{mc^2}\right)\mathbf{v}\,\mathbf{F}.\mathbf{v} \tag{2.21}
$$

This is an interesting result, which shows us that the force in general is **not proportional** to acceleration in the special theory of relativity. One can observe that in general, acceleration is not in the direction of force in the special theory of relativity. As a special case, when the velocity is perpendicular to the force, then the second term vanishes and the acceleration becomes parallel to the force.

2.3   Relativistic Energy: Mass-Energy Equivalence ($E = mc^2$)

We have already discussed that when a body of rest mass, m_0, moves with a velocity **v** with respect to a stationary observer, its relativistic mass is:

$$m = \frac{m_0}{\sqrt{1 - v^2/c^2}}. \tag{2.22}$$

As can be foreseen, the variation of mass with velocity will have its consequences on the concept of energy. Now, let the same body be acted upon by an external force **F** in the direction of the velocity of the body. According to Newton's second law of motion, force is the rate of change of momentum, and hence:

$$F = \frac{d}{dt}(mv) = m\frac{dv}{dt} + v\frac{dm}{dt}. \tag{2.23}$$

By the work-energy principle, the work done by the force in making a displacement of the body, dx, is the change in its kinetic energy:

$$\begin{aligned}
dT = Fdx &= m\frac{dv}{dt}\,dx + v\frac{dm}{dt}\,dx \\
&= m\frac{dx}{dt}\,dv + v\frac{dx}{dt}\,dm \\
&= mvdv + v^2dm \quad \left(\because \frac{dx}{dt} = v\right).
\end{aligned} \tag{2.24}$$

Now, differentiating Eq. (2.22):

$$\begin{aligned}
dm &= m_0(-1/2)\left(1 - v^2/c^2\right)^{-3/2}\left(-2vdv/c^2\right) \\
&= \frac{m_0}{c^2}\frac{vdv}{\left(1 - v^2/c^2\right)^{3/2}}.
\end{aligned} \tag{2.25}$$

Substituting $m_0 = [\sqrt{1 - v^2/c^2}]m$ in the above equation, dm becomes:

$$dm = \frac{mvdv}{(c^2 - v^2)}$$

$$\Rightarrow mvdv = (c^2 - v^2)dm.$$

Substituting this in Eq. (2.24):

$$dT = (c^2 - v^2)dm + v^2 dm$$
$$= c^2 dm \qquad (2.26)$$

If the rest mass of the body is m_0 and its mass is m, when the velocity reaches v, the required kinetic energy is:

$$T = \int dT = \int_{m_0}^{m} c^2 dm$$
$$= c^2(m - m_0)$$
$$\Rightarrow \boxed{T = mc^2 - m_0 c^2}. \qquad (2.27)$$

This is the expression for *"relativistic kinetic energy"*. This means the kinetic energy of the body is c^2 times the change in the relativistic mass of the body over its rest mass.

Let us now define the *energy of the body due to its inertia*, or *the rest energy*, or the *proper energy*, as:

$$E_0 = m_0 c^2 \qquad (2.28)$$

The total energy of the body is:

E = Kinetic energy + Rest energy = $T + E_0 = (m - m_0)c^2 + m_0 c^2$

$$\Rightarrow \boxed{E = mc^2} \qquad (2.29)$$

This is the famous **Einstein's mass-energy equivalence relation**. This energy, E is the *relativistic energy* or the *total energy* of a particle having *relativistic mass, m*.

The classical correspondence of the above equation can be shown by using the expression for relativistic kinetic energy and then, going to the non-relativistic velocities to show that we retrieve the expression for kinetic energy used in classical mechanics. The expression for relativistic kinetic energy is:

$$T = E - E_0 = (m - m_0)c^2$$
$$= m_0 c^2 (1 - v^2/c^2)^{-1/2} - m_0 c^2$$
$$\Rightarrow T = m_0 c^2 \left[1 + \frac{1}{2} \cdot \frac{v^2}{c^2} + \frac{3}{8} \cdot \frac{v^4}{c^4} + \ldots \ldots \right] - m_0 c^2 \quad \text{(using Binomial theorem)}$$

In the non-relativistic limit of $v/c << 1$, the higher-order terms can be neglected. Keeping up to $O(v^2/c^2)$, we get:

$$T = m_0 c^2 \left[1 + \frac{1}{2} \cdot \frac{v^2}{c^2} \right] - m_0 c^2$$

$$= \frac{1}{2} m_0 v^2. \tag{2.30}$$

This is the classical expression for kinetic energy.

There are several experimental observations to validate the mass-energy conversion.

- For example, in the case of electron-positron annihilation:

$$e^+ e^- \rightarrow \gamma \gamma$$

  When the rest mass of the left-hand side is finite, the right-hand side is zero, and this is an allowed reaction. Here, mass is converted to energy. The reverse interaction is also observed, meaning—energy can also be converted to mass.
- Further, the fine structure of spectral lines is explained using the relativistic variation of mass.
- In the Compton scattering experiment, the wavelength of a scattered photon from an atomic electron, which recoils in the process, is explained using energy-momentum conservation, considering the mass-energy interconversion relationship.
- In nuclear fission, energy is released. On August 6, 1945, the Hiroshima atomic bomb, *Little Boy* used 700 milligrams of highly-enriched fissile Uranium(235) to generate 6.3×10^{13} Joules of energy, which is equivalent to 15,000 tons of TNT.

2.3.1 Expression for Total Energy: Relativistic Case

If a particle of rest mass m_0 moves with a velocity v, its relativistic mass is m, and the total energy associated with the particle is the sum of its rest energy and the kinetic energy. The total energy is given by:

$$E = mc^2 = \gamma m_0 c^2 \tag{2.31}$$

Similarly, the relativistic momentum of the particle is:

$$\mathbf{p} = m\mathbf{v} = \gamma m_0 \mathbf{v}$$

$$\Rightarrow \mathbf{p}c = \gamma m_0 vc \tag{2.32}$$

Now, squaring both equations and subtracting the second equation from the first, one obtains:

$$E^2 - p^2c^2 = (\gamma m_0)^2 c^4 - (\gamma m_0)^2 v^2 c^2$$
$$= (\gamma m_0)^2 c^2 \left[c^2 - v^2 \right]$$
$$= (\gamma m_0)^2 c^4 \left[1 - v^2/c^2 \right]$$
$$= m_0^2 c^4$$

$$\Rightarrow \boxed{E^2 = p^2c^2 + m_0^2 c^4} \tag{2.33}$$

This is the expression for relativistic energy involving relativistic momentum and rest mass. In the subsequent chapters, we shall show that this expression satisfies the conservation of energy-momentum.

Example 2.3 Show that $E^2 = p^2c^2 + m_0^2 c^4$ can be reduced to the classical formula for kinetic energy, $T = \frac{p^2}{2m}$

Solution 2.3 Kinetic energy = total energy − rest energy

$$\Rightarrow T = E - m_0 c^2$$
$$= \sqrt{(p^2c^2 + m_0^2 c^4} - m_0 c^2$$
$$= m_0 c^2 \left[\left(1 + \frac{p^2}{m_0^2 c^2} \right)^{1/2} - 1 \right]$$

For non-relativistic case, $v << c$ i.e., $p << m_0 c$, then:

$$T = m_0 c^2 \left[\left(1 + \frac{1}{2} \frac{p^2}{m_0^2 c^2} \right) - 1 \right]$$
$$= \frac{p^2}{2m_0}$$

Further, for $v << c$, $m = m_0$. Hence, $T = \frac{p^2}{2m}$ is the classical formula for kinetic energy.

Example 2.4 Particles with zero rest mass: energy and Momentum

Solution 2.4 A particle with zero rest mass is called a massless particle. This is a non-classical concept, contrary to relativity, where physical properties like energy, momentum *etc.*, for a massless particle are well-defined. Let us now look into the requirements for the existence of a massless particle within the framework of relativity. The relativistic energy E of a particle with rest mass m_0 and momentum p, moving with velocity v are given by:

$$E = \sqrt{p^2 c^2 + m_0^2 c^4}, \text{ also } E = mc^2 = \gamma m_0 c^2 = \frac{m_0 c^2}{\sqrt{1 - v^2/c^2}}, \text{ and }$$

$$p = mv = \frac{m_0 v}{\sqrt{1 - v^2/c^2}}.$$

- When $m_0 = 0$, $E = pc$ or $p = E/c$.

  But $p = mv = v\frac{E}{c^2}$, hence $\frac{Ev}{c^2} = \frac{E}{c}$. This indicates $v = c$. *A massless particle travels at the speed of light.*

- If m is the relativistic mass of such a particle, like a photon having zero rest mass, then $E = mc^2$ or $m = E/c^2$ and $p = mc = E/c$.

  For a photon of frequency v, $E = hv$, where h is Planck's constant. Therefore, for a photon, *the relativistic mass, m and the momentum p are given by:*
  $m = \frac{E}{c^2} = \frac{hv}{c^2}$ and $p = \frac{E}{c} = \frac{hv}{c} = \frac{h}{\lambda}.$

- When $m_0 = 0$ and $v << c$, then $E = m_0 c^2$ and $p = m_0 v :\Rightarrow E = p = 0$.

  A massless particle in the non-relativistic limit has no energy and momentum.

- When $m_0 = 0$ and $v = c :\Rightarrow E$ and p are indeterminate.

  Zero mass particles with physical values of energy and momentum can only exist if they travel at the speed of light, like photons and neutrinos.

2.4   Lorentz Transformation of Energy and Momentum

As shown in Fig. 1.2, the inertial frame S' is moving away from the inertial frame S with a velocity v along the common $X - X'$ coinciding axes with their origins O, and O' being the same at $t = t' = 0$. Let the particle velocity as measured in S and S' be u and u', respectively. Using the formula for relativistic addition of velocities:

$$u' = \frac{u - v}{1 - uv/c^2} \tag{2.34}$$

In the stationary frame S,

$$E = mc^2 = \frac{m_0 c^2}{\sqrt{1 - u^2/c^2}} \tag{2.35}$$

and

$$p = mu = \frac{m_0 u}{\sqrt{1 - u^2/c^2}}. \tag{2.36}$$

In S'-frame, the corresponding quantities are:

$$E' = m'c^2 = \frac{m_0 c^2}{\sqrt{1 - u'^2/c^2}} \qquad (2.37)$$

and

$$p' = m'u' = \frac{m_0 u'}{\sqrt{1 - u'^2/c^2}}. \qquad (2.38)$$

Now, let us first evaluate $(1 - u'^2/c^2)^{-1/2}$ in trems of u using the velocity transformation formula.

$$\begin{aligned}
(1 - u'^2/c^2)^{-1/2} &= \left[1 - \frac{1}{c^2} \frac{(u-v)^2}{(1 - uv/c^2)^2} \right]^{-1/2} \\[2mm]
&= \left[\frac{(1 - uv/c^2)^2 - \frac{1}{c^2}(u-v)^2}{(1 - uv/c^2)^2} \right]^{-1/2} \\[2mm]
&= \left[\frac{(1 - u^2/c^2) - \frac{v^2}{c^2}(1 - u^2/c^2)}{(1 - uv/c^2)^2} \right]^{-1/2} \\[2mm]
&= \left[\frac{(1 - u^2/c^2)(1 - v^2/c^2)}{(1 - uv/c^2)^2} \right]^{-1/2} \\[2mm]
&= \frac{(1 - uv/c^2)}{\left[(1 - u^2/c^2)(1 - v^2/c^2) \right]^{1/2}} \qquad (2.39)
\end{aligned}$$

Using this in Eq. (2.37):

$$\begin{aligned}
E' = m'c^2 &= \frac{m_0 c^2}{\sqrt{1 - u'^2/c^2}} \\[2mm]
&= m_0 c^2 \frac{(1 - uv/c^2)}{\left[(1 - u^2/c^2)(1 - v^2/c^2) \right]^{1/2}} \\[2mm]
&= \frac{1}{(1 - v^2/c^2)^{1/2}} \left[\frac{m_0 c^2}{(1 - u^2/c^2)^{1/2}} - \frac{m_0 u v}{(1 - u^2/c^2)^{1/2}} \right] \\[2mm]
&= \frac{1}{(1 - v^2/c^2)^{1/2}} (E - pv) \quad \text{by using Eqs. (2.35) and (2.36))} \\[2mm]
\Rightarrow \quad &\boxed{E' = \gamma(E - pv)} \qquad (2.40)
\end{aligned}$$

Using Eq. (2.39) in Eq. (2.38) for momentum, we get:

$$p' = m'u' = \frac{m_0 u'}{\sqrt{1 - u'^2/c^2}}$$

$$= m_0 \frac{u - v}{1 - uv/c^2} \times \frac{(1 - uv/c^2)}{\left[(1 - u^2/c^2)(1 - v^2/c^2)\right]^{1/2}}$$

$$= \frac{1}{(1 - v^2/c^2)^{1/2}} \times \left[\frac{m_0 u}{(1 - u^2/c^2)^{1/2}} - \frac{m_0 v}{(1 - u^2/c^2)^{1/2}}\right]$$

$$\Rightarrow \boxed{p' = \gamma(p - Ev/c^2)}. \tag{2.41}$$

Equations (2.40) and (2.41) are the Lorentz transformation equations for energy and momentum. The inverse Lorentz transformation equations for energy and momentum are given by changing $v \rightarrow -v$ and making primed to unprimed and vice versa.

$$\boxed{\begin{aligned} E &= \gamma(E' + pv) \\ p &= \gamma(p' + Ev/c^2) \end{aligned}} \tag{2.42}$$

It should be noted here that the above transformations apply to the x-component of the momentum. The transverse components of the momentum (if the relative motion between the inertial frames is along the X-$axis$) are:

$$p_y = p'_y \text{ and } p_z = p'_z.$$

As the transformation equations for E and p are linear (i.e., no involvement of higher powers in E and p), these equations are equally valid for the total energy and momentum of a system of particles. In the latter case, for the direct transformations:

$$\boxed{\begin{aligned} E'_{total} &= \gamma(E_{total} - vp_{total}) \\ p'_{total} &= \gamma\left(p_{total} - \frac{vE_{total}}{c^2}\right) \end{aligned}} \tag{2.43}$$

Example 2.5 Prove that $E^2 - p^2c^2 = m_0^2c^4$ is Lorentz invariant

Solution 2.5 Following the above discussions, it is clear from the relativistic formula for energy, $E^2 = p^2c^2 + m_0^2c^4$, that $E^2 - p^2c^2 = m_0^2c^4$ is Lorentz invariant. Let us show it explicitly using the transformation equations for energy and momentum.

$$E' = \gamma(E - pv)$$

$$p' = \gamma(p - Ev/c^2)$$

Now

$$
\begin{aligned}
E'^2 - p'^2 c^2 &= \gamma^2 \left[E^2 + p^2 v^2 - 2Epv - \left(p^2 + \frac{E^2 v^2}{c^4} - \frac{2Epv}{c^2} \right) c^2 \right] \\
&= \gamma^2 \left[E^2 + p^2 v^2 - p^2 c^2 - \frac{E^2 v^2}{c^2} \right] \\
&= \gamma^2 \left[E^2 \left(1 - \frac{v^2}{c^2} \right) - p^2 c^2 \left(1 - \frac{v^2}{c^2} \right) \right] \\
&= \gamma^2 \left(1 - \frac{v^2}{c^2} \right) \left[E^2 - p^2 c^2 \right] \\
&= E^2 - p^2 c^2,
\end{aligned}
$$

which is Lotentz invariant.

Choice of System of Units

3

On July 23, 1983, Air Canada Flight 143, a domestic passenger flight between Montreal and Edmonton, was in danger because it ran out of fuel when it was cruising at an altitude of 41,000 feet. With 69 people on board, the pilot managed to land the plane safely, as he was an experienced glider pilot. The fuel ran out as 22,300 pounds of aviation fuel was filled instead of 22,300 kilograms (49,163 lb). In aviation history, it is known as *"Gimli Glider"*. This is a beautiful example of the importance of using proper units.

Quantities that are dimensionless in one system of units may have dimensions in other systems. According to Max Planck, *"the fact that when a definite physical quantity is measured in two systems of units it has not only different numerical values, but also different dimensions has often been interpreted as an inconsistency that demands explanation, and has given rise to the question of the 'real' dimensions of a physical quantity."* As a reminder to the readers, any book on electrodynamics deals with various systems of units, like the International System of Units (Système International d'Unités) (SI), Heaviside-Lorentz, Gaussian, electrostatic and electromagnetic, *etc.*, leading to different forms of physical equations. A nautical mile (equivalent to 1852 meters) is the unit of distance used in sea or aviation navigation, which is equivalent to one *arc minute of a geodesic* (a great circle). The derived unit of speed is *knot*, one nautical mile per hour. The speed of an aircraft is sometimes measured in *Mach* to determine if it is subsonic or supersonic. These are some of the unconventional uses demanding proper conversions to have a realization in day-to-day life. Going inline with this discussion, while making a transition from classical physics to subatomic physics, where new emergent phenomona come in, we need to deal with a special system of units, called *natural units*. This is the system of units usually used in high-energy physics while dealing with the kinematics of subatomic particles, where one chooses Planck's constant ($\hbar$) and the speed of light in vacuum (c) to be one. In the following section, we shall discuss natural units with the underlying physics, while dealing with some of the more frequently used observables in high-energy physics.

3.1   Why Natural Units in High Energy Physics?

In high-energy nuclear and particle physics, one typically deals with subatomic particles, which move at nearly the speed of light in vacuum, and hence, the special theory of relativity has an important place, making the classical Newtonian physics of no practical use. For instance, although one deals with elementary particles like quarks and gluons (collectively called partons), for simplicity, if one takes a proton, the associated rest energy is:

$$E_p = m_p c^2 = (1.6 \times 10^{-27}\, kg) \times (3 \times 10^8\, m/sec)^2 = 1.503 \times 10^{-10}\, Joule. \tag{3.1}$$

This is a very tiny number and one has to always deal with negative exponents of several orders while discussing mass, energy, *etc.* of subatomic particles. If one takes *electron-volt*, which is the change in energy of an electron, when it traverses through a potential difference of one volt, taking electron charge as $e = 1.602 \times 10^{-19}\, C$ (Coulomb) and $1V = 1\, Joule/C$, the conversion is therefore:

$$1\, eV = (1.602 \times 10^{-19}\, C) \times 1\, Joule/C = 1.602 \times 10^{-19}\, Joule.$$

Using this conversion, the proton rest energy is:

$$E_p = 1.503 \times 10^{-10}\, Joule = \frac{1.503 \times 10^{-10}}{1.602 \times 10^{-19}}\, eV = 0.938 \times 10^9\, eV$$

$$= 938\, MeV. \tag{3.2}$$

Similarly, our known electron, which is around 1836 times smaller in mass as compared to a proton, has a rest mass of $9.11 \times 10^{-31}\, kg$. This is equivalent to $E_e = 0.511\, MeV$. The Higgs boson mass or the rest energy is around 125 GeV. As in high-energy physics, physicists use MeV and GeV as the unit of mass; the right way to say this is—we need to divide c^2 with energy to get the mass. This follows from the definition of rest energy, $E_{rest} = m_0 c^2$. The best way to use it is if we enter a system of units, where we can, in fact, ignore c by setting $c = 1$. This will follow as a part of the system of natural units.

As we proceed, we shall discover that the formulae which describe the kinematics of such subatomic particles will involve Planck's constant and the speed of light in vacuum. We know, the length element in 4-dimensional Minkowski space is given by

$$x^2 = c^2 t^2 - x_1^2 - x_2^2 - x_3^2 \tag{3.3}$$

and the relativistic energy of a particle is

$$E = \sqrt{\mathbf{p}^2 c^2 + m_0^2 c^4}. \tag{3.4}$$

The velocity of light "c" appears directly in these frequently used formulas and their alikes. Furthermore, de Broglie relation between 4-momentum and wave vector of a particle is

$$p^\mu = \hbar k^\mu$$
$$\Rightarrow p = \hbar k, \tag{3.5}$$

where $p = (\frac{E}{c}, \mathbf{p})$, $k = (\frac{\omega}{c}, \mathbf{k})$. Here ω is the angular frequency and $\mathbf{k}$ is the wave vector. Component-wise one writes

$$\left(\frac{E}{c}, \mathbf{p} \right) = \hbar \left(\frac{\omega}{c}, \mathbf{k} \right) \tag{3.6}$$

In a system of units with $c = 1$, this equation takes a simpler form and can be written as

$$E = \hbar \omega \tag{3.7}$$

$$\mathbf{p} = \hbar \mathbf{k} \tag{3.8}$$

This Planck's constant ($\hbar$) relates the wavelength of a particle wave to the momentum associated with the particle

$$\lambda = 2\pi \hbar / |\mathbf{p}| \tag{3.9}$$

If we choose a system of units in which

$$\boxed{\hbar = c = 1},$$

where $\hbar = \frac{h}{2\pi} = 1.055 \times 10^{-34}$ *Joule. sec*: unit of action/angular momentum (ML^2/T).

$c = 2.998 \times 10^8$ *meter. sec*$^{-1}$: unit of velocity, the velocity of light in vacuum (L/T).

Now the relativistic formula for energy, which connects its rest mass and momentum, is given by

$$E^2 = \mathbf{p}^2 c^2 + m_0^2 c^4. \tag{3.10}$$

In this new system of units (called natural units), which is more popularly used in high-energy (particle) physics, it becomes

$$\boxed{E^2 = \mathbf{p}^2 + m_0^2} \tag{3.11}$$

We can define a system of units completely if we specify the unit of energy (ML^2/T^2). The equivalence of mass and energy allows us to use the energy unit to express all the observables. In this scheme, momentum $(\mathbf{p}c)$, energy (E), mass $(m_0 c^2)$, inverse of time $(\hbar/t)$, and inverse of length $(\hbar c/l)$ all have the same dimensions, which we take to be of energy. The unit of energy is usually taken as electron volt (eV). One electron volt (eV) is the amount of energy gained (ΔU) by one electron of charge e, if allowed inside a potential difference (ΔV) of one volt, following $\Delta U = e\Delta V$. It is interesting to note here that from atomic energy in the length scale of few angstrom (1 Å$= 10^{-10}$m $= 0.1$ nm), when we move to nuclear energy scale with a typical length scale of few fermi ($1\,fm = 10^{-15}$m), in the energy units, we travel from few electron volt (eV) to few GeV. In particle physics, looking into the length scale we explore, more often the unit of energy used is MeV (= million/mega electron volt) or GeV (1 GeV $= 10^9$ eV). The rest masses of the electron and proton are 0.5 MeV and 938.3 MeV, respectively. This choice of energy units is motivated by the rest mass of the proton $\sim$ 1 GeV and thus gives a sense of physical intuition/realisation. This gives rise to mass (m_0), momentum $(m_0 c)$, and energy $(m_0 c^2)$ in GeV. Length $(\frac{\hbar}{m_0 c})$ and time $(\frac{\hbar}{m_0 c^2})$ in GeV^{-1}.

3.2　　Natural Units

In natural units, we take $\hbar = c = 1$ and, for example, we derive some of the units in the unit of energy as follows. We know

$$\hbar = 1.054 \times 10^{-34} \ J.sec$$

$$\Rightarrow 1\,sec = \frac{\hbar}{1.054 \times 10^{-34} \ J}$$

$$1\,J = \frac{1}{1.602 \times 10^{-19}} \ eV$$

$$\Rightarrow 1\,sec = \frac{1.602 \times 10^{-19}}{1.054 \times 10^{-34}} \ eV^{-1}$$

$$\Rightarrow \boxed{1\,sec \;=\; 1.52 \times 10^{24}\,GeV^{-1}} \tag{3.12}$$

Let us now convert the unit of length, one meter, to energy units.

$$c \;=\; 2.998 \times 10^{8}\,m/sec$$

$$\Rightarrow 1\,m \;=\; \frac{c.sec}{2.998 \times 10^{8}} \;=\; \frac{1.52 \times 10^{24}\,GeV^{-1}}{2.998 \times 10^{8}}\;(\text{using Eq. (3.12)})$$

$$\Rightarrow \boxed{1\,meter \;=\; 5.07 \times 10^{15}\,GeV^{-1}} \tag{3.13}$$

Some useful conversions are listed below.

$$1\,fermi \;\equiv\; 1\,fm \;=\; 10^{-13}\,cm \;=\; 10^{-15}m$$

$$\Rightarrow \boxed{1\,fm \;=\; 5.07\,GeV^{-1}}$$

$$\boxed{1\,fm \;=\; 3.33 \times 10^{-24}\,sec}$$

$$\boxed{197.3\,MeV \;=\; 1\,fm^{-1}}$$

Note: $1\,\text{TeV} = 10^{3}\,\text{GeV} = 10^{6}\,\text{MeV} = 10^{9}\,\text{KeV} = 10^{12}\,\text{eV}$.
The following conversions may be useful in the subsequent discussions:

$$\boxed{\hbar c \;=\; 197.3\,\text{MeV. fm} \simeq 0.2\,\text{GeV. fm}}$$
$$\boxed{(\hbar c)^{2} \;=\; 0.389\,(\text{GeV})^{2}\,\text{mb}}$$

Cross Sections

Cross sections are measured in millibarns ($1\,\text{mb} = 10^{-31}\,m^{2} = 10^{-27}\,cm^{2} = 0.1\,fm^{2}$), or micro-barns ($1\,\mu b = 0.001\,mb$), nano-barns ($1\,\text{nb} = 0.001\,\mu b$) or pico-barns ($1\,\text{pb} = 0.001\,\text{nb}$). Cross sections have the dimension of energy squared-inverse in natural units.

$$[\sigma] = [L]^{2} = [M]^{-2}$$

$$1\,fm = \frac{1}{197.3}\,MeV^{-1} = 5.068\,GeV^{-1}$$

$$\Rightarrow 1\,fm^{2} = 25.7\,GeV^{-2}$$

$$1\,barn = 10^{-24}\,cm^{2} = 100\,fm^{2}$$

Hence, it follows that

$$\boxed{1 \text{ GeV}^{-2} = 0.3894 \text{ mb}, \quad 1 \text{ mb} = 2.568 \text{ GeV}^{-2}, \quad 1 \text{ fm}^2 = 10 \text{ mb}}.$$

Equivalently, using the known values of pion and proton masses m_π and m_p, one has

$$\frac{1}{m_\pi^2} = 19.987 \text{ mb}$$

$$\frac{1}{m_p^2} = 0.44232 \text{ mb} \tag{3.14}$$

These are very frequently used by high-energy physicists. The additional advantage of using natural units in high energy physics is that we deal with strong interaction, whose lifetime $\sim 10^{-24} sec$, the decay length of a particle can be better expressed in terms of fermi.

Example 3.1 Hadronic Cross-section

Solution 3.1 A typical hadronic cross-section is of the order of

$$\sigma \simeq \lambda_\pi^2 \simeq \frac{1}{m_\pi^2} \simeq \frac{1}{(140)^2} MeV^{-2} \tag{3.15}$$

If σ is to be expressed in fm^2, we need to multiply a combination of $\hbar$ and c with units $MeV^2.fm^2$, to bring out proper units.
This is

$$\hbar^2 c^2 = (197.3)^2 MeV^2.fm^2$$

and $\sigma = \frac{\hbar^2 c^2}{(140)^2 \, MeV^2} \simeq 2 \; fm^2 \simeq 20 \; mb.$

Example 3.2 Kelvin to MeV conversion

Solution 3.2 Recall, in statistical thermodynamics, one uses the equation $E \sim k_B T$, which relates energy with temperature through the Boltzmann constant, $k_B = 8.617 \times 10^{-5} \, eV/K$, which equals $1.38 \times 10^{-23} \, kg.m^2.s^{-2}.K^{-1}$. In natural units $k_B = 1$. This implies:

$$8.617 \times 10^{-5} \, eV/K = 1$$

$$\Rightarrow 1 \, K = 8.617 \times 10^{-5} \, eV$$

$$\Rightarrow 1 \, K = 8.617 \times 10^{-11} \, MeV \tag{3.16}$$

$$\Rightarrow \boxed{1\,MeV = 1.1605 \times 10^{10}\,K}. \tag{3.17}$$

To have a feeling of the temperature achieved in relativistic heavy-ion collisions, let us do the following exercise. For example, room temperature is around $300\,K = 300 \times 8.62 \times 10^{-11}\,MeV \simeq 2.585 \times 10^{-8}\,MeV$. The QCD phase transition temperature is estimated to be around $156\,MeV$. This is equivalent to:

$T = 170 \times (1.1605 \times 10^{10})\,K = 1.97 \times 10^{12}\,K$. The core of the Sun's temperature is around 15.7 million degrees Kelvin, i.e. $1.57 \times 10^{7}\,K$. Hence, the temperature achieved in relativistic heavy-ion collisions is estimated to be around 10^{5} times the core of the Sun's temperature, the highest temperature achieved in laboratories.

Example 3.3 Magnetic field produced in heavy-ion collisions

Solution 3.3 Magnetic fields produced at the early stages of Au-Au collisions at RHIC with a center-of-mass energy $\sqrt{s_{NN}} = 200\,GeV$ and impact parameter $b = 4\,fm$ are about $eB \approx 1.3\,m_\pi^2$ [2]. This is created because of the relativistic motion of the charged spectators as a constituent of colliding heavy ions. This is about four orders of magnitude higher as compared to that observed on the surface of a magnetar. As a side note, this is the highest magnetic field achieved in the laboratory. This is because making steady fields stronger than 4.5×10^5 Gauss in the laboratory is impossible because magnetic stresses of such fields usually exceed the tensile strength of terrestrial materials. As one notices here, in high energy physics, we use magnetic field strength in the units of m_π^2, whereas classically the units are either Gauss or Tesla (1 Tesla = 10^4 Gauss). Now we proceed to show the magnetic field strength in natural units.

Lorentz force on a charge q moving with a velocity $\mathbf{v}$ in a magnetic field $\mathbf{B}$, is given by

$$\mathbf{F} = q(\mathbf{v} \times \mathbf{B})$$

$$F = qvB \quad \text{(taking the angle as } 90^0 \text{ for simplicity)}$$

$$\Rightarrow B = F/qv = \frac{N}{C.m/s} = kg.\frac{m}{s^2} \times \frac{s}{C.m}$$

$$\Rightarrow Tesla = \frac{kg}{C.s} \quad \text{[In S.I. units]} \tag{3.18}$$

Now, let us estimate the energy equivalent of 1 kg in natural units. We start with Einstein's equation for mass-energy equivalence:

$$E = mc^2$$

$$= (1\,kg)\,c^2 = 1\,kg \times (2.998 \times 10^8\,m/s)^2$$

$$= 8.987 \times 10^{16}\,kg.m^2/s^2 = 8.987 \times 10^{16}\,J \tag{3.19}$$

Given $1\,eV = 1.6022 \times 10^{-19}\,J$, $1\,J = 6.2415 \times 10^{18}\,eV$.

Hence, $1\,kg = 8.987 \times 10^{16}\,J \times 6.2415 \times 10^{18}\,eV/J = 5.6095 \times 10^{35}\,eV$.

$$\Rightarrow \boxed{1\,kg = 5.6095 \times 10^{26}\,GeV}. \tag{3.20}$$

Now, let us proceed to convert Coulomb to natural units.

The fine structure constant is given by, $\alpha = \frac{e^2}{4\pi\epsilon_0 \hbar c} \approx \frac{1}{137}$.

In natural units $\hbar = c = 1$ and $\epsilon_0 = 1$ (using Heaviside-Lorentz unit).

Hence,

$$e^2 = \alpha.4\pi\epsilon_0\hbar c = 4\pi\alpha = \frac{4\pi}{137}.$$

$$\Rightarrow \boxed{e = \sqrt{\frac{4\pi}{137}} \approx 0.3028}. \tag{3.21}$$

This is the value of the elementary charge in natural-Heaviside-Lorentz units.

Recall, $1\,e = 1.602 \times 10^{-19}\,C$.

$$\Rightarrow 1\,C \approx 6.242 \times 10^{18}\,e = 6.242 \times 10^{18} \times 0.3028.$$

$\Rightarrow \boxed{1\,C \approx 1.89 \times 10^{18}}$, which is dimensionless in natural-Heaviside-Lorentz units.

Now,

$$1\,Tesla = \frac{1\,kg}{1\,C \times 1\,s} = \frac{5.6095 \times 10^{26}\,GeV}{(1.89 \times 10^{18}) \times (1.52 \times 10^{24})\,GeV^{-1}}$$

$$\Rightarrow 1\,Tesla = 1.9526 \times 10^{-16}\,GeV^2.$$

As $1\,Tesla = 10^4\,Gauss$, $\boxed{1\,Gauss = 1.9526 \times 10^{-20}\,GeV^2}$

$$\Rightarrow \boxed{1\,MeV^2 = 0.5128 \times 10^{14}\,Gauss}.$$

Now,

$m_\pi^2 (140\,MeV)^2 = 19,600\,MeV^2 = 19,600 \times (0.5128 \times 10^{14})\,Gauss = 1.005 \times 10^{18}\,Gauss \approx 10^{18}\,Gauss$.

$$\Rightarrow \boxed{m_\pi^2 = 10^{18}\,Gauss}.$$

Earth's magnetic field strength is around 25–$65\,\mu$T or 0.25 to $0.65\,Gauss$, whereas a typical magnetic field produced in heavy-ion collisions is around

10^{18} *Gauss*, the highest produced in the laboratory. Note here that this magnetic field is highly transient in nature and decays within a few fermi time.

Example 3.4 The shear viscosity to entropy ratio

Solution 3.4 One of the most useful observables in the characterization of the medium formed in heavy-ion collisions at relativistic energies is the shear viscosity to entropy density ratio (η/s). When compared to the corresponding estimate from AdS/CFT (anti-de sitter/ conformal field theory), which gives a value of $\eta/s = 1/4\pi \approx 0.08$, one finds that the matter created in heavy-ion collisions (Au+Au at $\sqrt{s_{NN}} = 200\,\text{GeV}$) at the Relativistic Heavy-Ion Collider (RHIC) at the Brookhaven National Laboratory, USA is *"the most perfect fluid"* with the lowest value of η/s [3].

In SI units, the shear viscosity is $\eta = \frac{shear\ stress}{velocity\ gradient}$, which is the ratio of force per unit area (pressure in Pascal) and the inverse of time. The unit of η is Pascal. sec. This is nothing but $\frac{kg.m/s^2}{m^2}/(1/s) = \frac{kg}{m.s}$.

In natural units, $[\eta] = [\frac{kg}{m.s}] = [\frac{E}{E^{-1}.E^{-1}}] = [E^3]$ i.e., (GeV^3).

Entropy density, $s = \frac{entropy}{volume} = \frac{S}{V}$. Entropy, $S = \frac{Q}{T}$, both are in energy units. Hence, S is dimensionless in natural units. As volume has the dimension of $1/E^3$, entropy density is expressed in GeV^3 in natural units. This means η/s is dimensionless in natural units.

Example 3.5 The de Broglie Wavelength

Solution 3.5 The de Broglie wavelength associated with a 1 GeV photon is given by

$$\lambda_\gamma^{dB} = \frac{h}{p} = \frac{2\pi\hbar c}{E} = 2\pi \ GeV^{-1} \tag{3.22}$$

$$= 2\pi \times 0.1975 \times 10^{-15} \ m$$

$$= 0.395\pi \ fm.$$

Example 3.6 The reduced compton wavelength

Solution 3.6 It is the wavelength associated with the matter particle and thus represents the length scale at which relativistic quantum field theory becomes crucial for its accurate description. For example, the Compton wavelength of an electron is the characteristic length scale of Quantum Electrodynamics (QED). The Compton wavelength of a particle of mass m is defined as

$$\lambda = \frac{h}{mc} \tag{3.23}$$

The reduced Compton wavelength is given by

$$\lambda/(2\pi) \equiv \lambda_C = \frac{\hbar c}{mc^2} = \frac{\hbar}{mc} \tag{3.24}$$

$$\Rightarrow \boxed{\lambda_C = \frac{1}{m}} \quad \text{(in   N.U.)}$$

In natural units, this takes a very simple form, being inversely proportional to the particle mass. Compton wavelength for a pion ($m_\pi \sim 140\, MeV$) is given by
$\lambda_C^\pi = \frac{1}{140\, MeV} = \frac{197}{140}\, fm = 1.411\, fm$ (using $1\, GeV^{-1} = 0.1975\, fm$
or $1\, MeV^{-1} = 197\, fm$).
For an electron, whose mass is $\sim 0.5\, MeV$,

$$\lambda_C^e = \frac{1}{0.5\, MeV} = \frac{197}{0.5}\, fm = 395\, fm.$$

For a proton,

$$\lambda_C^p \simeq 1\, GeV^{-1} = 0.1975\, fm.$$

Example 3.7   Classical electron radius

Solution 3.7 In theoretical models, where the electron is considered as an extended object, one estimates the classical electron radius. This is also known as the *Lorentz radius* or the *Thomson scattering length*. This is given by

$$r_e = \frac{1}{4\pi \epsilon_0} \frac{e^2}{mc^2}. \tag{3.25}$$

In Natural-Heaviside-Lorentz units, we have $\epsilon_0 = \mu_0 = c = 1$ and $e \simeq 0.3028$. Using these, we get,

$$r_e = \frac{e^2}{4\pi m}$$

$$= \frac{0.3028^2}{4\pi (0.5\, MeV)}$$

$$= \frac{0.3028^2 \times 197}{4\pi \times 0.5}\, fm$$

$$= 2.88\, fm. \tag{3.26}$$

Example 3.8 The Bohr radius

Solution 3.8 This is the radius of the lowest energy stable orbit of the atomic electron. The energy of the electron in a Hydrogen atom is given by,

$$E \simeq \frac{p^2}{2m} - \frac{\alpha}{r}.$$ (3.27)

The first term represents the kinetic energy, and the 2nd term the electrostatic potential. The momentum p scales like $\frac{1}{r}$ and hence

$$E \simeq \frac{1}{2mr^2} - \frac{\alpha}{r}.$$

Taking the stability condition, $\frac{dE}{dr} = 0$, one gets the Bohr radius:

$$\begin{aligned}
r_B &= \frac{1}{\alpha m} \\
&= \frac{137}{0.5 \, MeV} \\
&= \frac{137 \times 197}{0.5} \, fm \\
&= 0.541 \times 10^{-10} \, m \\
&= 0.541 \, \text{Å}.
\end{aligned}$$

$$(3.28)$$

3.3   Representations of the Field Operators

1. In general, while dealing with relativistic subatomic particles, one deals with 4-dimensional Minkowski space. The Lagrangian has the dimension of mass, i.e.
   $[Lagrangian] = [M]$.
   The Lagrangian density thus gets the following dimensions
   $[\mathcal{L}] = [M][L]^{-3} = [M]^4$.
2. The Hamiltonian and the Hamiltonian density have respective dimensions like the Lagrangian and the Lagrangian density.
3. The action, which has the unit of $\hbar$ is dimensionless in natural units.
4. For a spinor field, $[\psi] = [M]^{3/2}$, and for a scalar field, $[\phi] = [M]$.
5. For photon and massive vector fields, $[A_\mu] = [M]$.
6. In case of derivative or covariant derivative operators, $[\partial^\mu] = [D^\mu] = [M]$.
7. Using the dimensions used above, one can find the dimensions of any coupling constant appearing in the interaction Lagrangian density.

3.4   The Cosmological Connection

Usually in the context of cosmology, one takes
$\hbar = c = k_B = 1$, where k_B is Boltzmann constant.

$$\boxed{1\,GeV = 1.2 \times 10^{13}\,Kelvin} \tag{3.29}$$

The Gravitational constant is given by

$$\boxed{G = \frac{1}{M_p^2}}, \tag{3.30}$$

where Planck mass, M_p is given by

$$\boxed{M_p = 1.2 \times 10^{19}\,GeV}. \tag{3.31}$$

The corresponding Planck length is

$$\boxed{l_p = \frac{1}{M_p} = 1.6 \times 10^{-33}\,cm}. \tag{3.32}$$

To get the corresponding Planck time, one needs to multiply $1/c$ to get,

$$\boxed{t_p = 5.4 \times 10^{-44}\,sec}. \tag{3.33}$$

Using Eqs. 3.29 and 3.31, the Planck temperature is given by

$$\boxed{T_p = 1.4 \times 10^{32}\,Kelvin}. \tag{3.34}$$

Minkowski Space: Four-Dimensional Formulation

4

> *Space by itself, and time by itself, are doomed to fade away into mere shadows, and only a kind of union of the two will preserve an independent reality.*
>
> *— Hermann Minkowski*

Hermann Minkowski, one of the teachers of Albert Einstein at ETH Zurich, gave a theory of four-dimensional space-time, which is known as the *"Minkowski space-time"*—a geometrical interpretation of the special theory of relativity. Later, the Minkowski spacetime framework laid the mathematical foundation for Einstein's general theory of relativity. In this chapter, we plan to touch upon some of the aspects of four-dimensional space-time, which may be needed while covering the four-vector formalism to solve kinematic problems in high-energy physics. We shall, however, not go into the details of geometrical interpretations, leaving them aside.

The special theory of relativity, with its two postulates, namely the constancy of speed of light in vacuum and the invariance of all laws of physics in inertial frames (frames of reference connected by a constant velocity), uses the Lorentz transformation properties of space-time coordinates of an event in two inertial frames. Unlike in Galilean transformations, the space and time coordinates are interdependent—the temporal coordinate of one frame is mathematically dependent on the spatial and temporal coordinates of another system (recall the Lorentz transformation equation for the time coordinate). It then becomes obvious to treat space and time coordinates together on the same footing than treating them separately. It was Hermann Minkowski who came up with an elegant idea to treat time as another coordinate, similar to the three spatial coordinates, forming a four-dimensional coordinate system, called the *"Minkowski space"*.

Although, as we have already seen explicitly in the previously discussed chapters, the mathematical treatments of the special theory of relativity don't need the 4-

R. Sahoo, *Relativistic Kinematics*, Lecture Notes in Physics 1046, https://doi.org/10.1007/978-3-032-09510-7_4

vector formalism, it becomes a necessity while dealing with the general theory of relativity. The general theory of relativity deals with tensors, which are a generalization of 4-vectors. However, the 4-vector formalism has a mathematical elegance in dealing with the subject of the special theory of relativity and the kinematic problems, which we shall see in a while.

4.1   Some Useful Formulae

As we have already introduced the system of natural units, henceforth, we shall be using it for the ease of dealing with the kinematic problems in high-energy physics. Before we proceed, let us recall the following useful formulae and concepts, which we shall be using frequently.

1. c = speed of light in vacuum = 2.99×10^8 m/sec.
2. $\beta = v/c$, v being the velocity of the particle under consideration, β denotes the fraction of the particle velocity in comparison to the maximum attainable velocity, i.e. speed of light, c, in vacuum. The variation of β as a function of velocity is shown in the left panel of Fig. 4.1.
3. $E = m_0c^2 + T$, T is the kinetic energy = $\frac{1}{2}mv^2$, the energy associated with the motion of the particle. The first term, m_0c^2 is the energy associated with the inertia of the particle, i.e. the rest mass energy.
4. The Lorentz factor, $\gamma = \frac{1}{\sqrt{1-\beta^2}} = \frac{E}{m_0c^2}$. The variation of γ as a function of velocity is shown in the right panel of Fig. 4.1.
5. As $v \leq c$, $0 \leq \beta \leq 1$. As β approaches 1, γ approaches ∞.
6. For relativistic particles:
   v is close to c,
   β is close to 1,
   γ is very large,

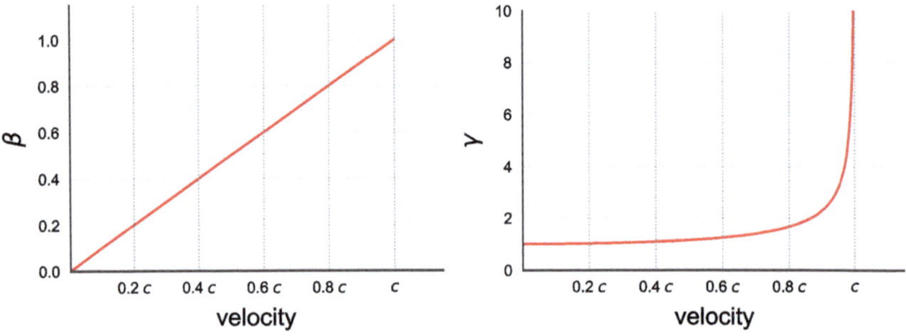

Fig. 4.1 The variation of β, the fraction of particle velocity in comparison to the maximum attainable velocity, c, as a function of the velocity of the particle (left). The variation of the Lorentz factor, γ as a function of velocity (right)

$$pc \gg m_0c^2,$$
$$T \gg m_0c^2.$$

7. Relativistic mass,

$$m = \frac{m_0}{\sqrt{1 - \beta^2}} = \gamma m_0. \tag{4.1}$$

8. Total energy of a particle obeying "on the mass shell" condition is,

$$E = \sqrt{p^2c^2 + m_0^2c^4} = \gamma m_0c^2 = \gamma m_0. \tag{4.2}$$

9. The particle momentum is,

$$p = \gamma m_0 v = \gamma m_0 \beta c = \gamma m_0 \beta. \tag{4.3}$$

10. The Lorentz transformation in natural units:

$$\boxed{\begin{aligned} x' &= \gamma(x - vt) = \gamma(x - \beta t), \\ t' &= \gamma\left(t - \frac{vx}{c^2}\right) = \gamma(t - \beta x), \\ y' &= y, \\ z' &= z. \end{aligned}} \tag{4.4}$$

with

$$\gamma = \frac{1}{\sqrt{1 - \frac{v^2}{c^2}}} = \frac{1}{\sqrt{1 - \beta^2}}.$$

The inverse Lorentz coordinate transformation is:

$$\boxed{\begin{aligned} x &= \gamma(x' + \beta t'), \\ t &= \gamma(t' + \beta x'), \\ y &= y', \\ z &= z'. \end{aligned}} \tag{4.5}$$

Example 4.1 Estimation of **the Lorentz factor, γ and β of a beam.**

Solution 4.1 Consider proton+proton collisions at a center-of-mass energy of $\sqrt{s} = 13$ TeV at the Large Hadron Collider, CERN.

$$m_0 c^2 = 938 \text{ MeV}$$

$$\gamma = \frac{E}{E_{\text{rest}}} = \frac{\gamma m_0 c^2}{m_0 c^2}$$

$$\Rightarrow \gamma = \frac{6.5 \times 10^6 \text{ MeV}}{938 \text{ MeV}} = 6929$$

Here, $E = \frac{\sqrt{s}}{2}$, is the beam energy. Now,

$$\gamma = \frac{1}{\sqrt{1 - \beta^2}}$$

$$\Rightarrow \beta = \sqrt{1 - 1/\gamma^2} = 0.9999999896$$

In order to achieve $\beta = 0.9999999996$, what will be the center-of-mass energy? A back calculation will give the beam energy, $E = 33.16$ TeV and hence the center-of-mass energy of 66.32 TeV. This tells about the power of relativistic energy and the associated relativistic mass– a change in the relativistic velocity at the eighth decimal place makes the relativistic mass so heavy that the required center-of-mass energy in this case becomes almost fivefold!

4.2   Four Vectors

In the four-dimensional Minkowski space, $A = (A_0, A_1, A_2, A_3)$ is called a *4-vector*, if the components, A_i transform under the Lorentz transformation in the same way as (ct, x, y, z). In other words, $A = (A_0, A_1, A_2, A_3)$ is defined as a *4-vector* if, under Lorentz transformation (inverse here):

$$A_0 = \gamma (A_0' + \beta A_1')$$
$$A_1 = \gamma (A_1' + \beta A_0')$$
$$A_2 = A_2'$$
$$A_3 = A_3' \tag{4.6}$$

Here, $A' \equiv (A_0', A_1', A_2', A_3')$ is a 4-vector. We are going to consider the zeroth component as the temporal component, and the first, second, and third components are the spatial components in this four-dimensional space. While using the above Lorentz transformation, we have assumed the conventional boost along the positive

$X - axis$ as shown in the Fig. 1.2. One can easily identify that we have replaced ct with A_0, x with A_1, y with A_2, and z with A_3. In addition, the last three components of the 4-vector constitute the standard 3-dimensional Euclidean space, and they transform like a usual 3-component vector under rotation in 3-dimensional space. Hence, in principle, *the definition of a 4-vector is that it must transform like* $(ct,\ x,\ y,\ z)$ *under Lorentz transformations and rotation*. Note that the coordinates referred to here are the intervals of space and time. Hence, $(ct,\ x,\ y,\ z)$ is essentially $(cdt,\ dx,\ dy,\ dz)$—just to get rid of any ambiguity.

Some Important Conventions and Notes
- We shall now work in the system of natural units, setting $c = 1$.
- A 4-vector will be denoted by an italic upper case or lower case letter, like A or a, while the usual 3-vector will be denoted as a bold-face letter, like $\mathbf{A}$ or $\mathbf{a}$.
- By convention, we shall use the first component of a 4-vector as the "temporal component", the other three as the spatial components.
- In some books, sometimes the first three are chosen as space, and the last one is the time component.
- While using contravariant and covariant notations for 4-vectors, we shall use the index μ, which runs from zero to three, where the zeroth component is the temporal component, and the first, second, and third components are the spatial components.
- A 4-vector is a 4-dimensional generalization of the 3-vector in 3-dimensional space, where the latter is invariant under a rotation. The 4-dimensional length element, as was shown earlier, is also invariant under Lorentz transformation.

The above transformation equations under a direct Lorentz transformation can be put in a matrix form as:

$$
\begin{bmatrix} A'_0 \\ A'_1 \\ A'_2 \\ A'_3 \end{bmatrix} = \begin{bmatrix} \gamma & -\beta\gamma & 0 & 0 \\ -\beta\gamma & \gamma & 0 & 0 \\ 0 & 0 & 1 & 0 \\ 0 & 0 & 0 & 1 \end{bmatrix} \begin{bmatrix} A_0 \\ A_1 \\ A_2 \\ A_3 \end{bmatrix} \tag{4.7}
$$

Now,

$$
(A'_0)^2 - (A'_1)^2 - (A'_2)^2 - (A'_3)^2
$$

$$
= \gamma^2 (A_0 - \beta A_1)^2 - \gamma^2 (A_1 - \beta A_0)^2 - A_2^2 - A_3^2
$$

$$
= \gamma^2 \left(A_0^2 + \beta^2 A_1^2 - 2\beta A_0 A_1 - A_1^2 - \beta^2 A_0^2 + 2\beta A_0 A_1 \right) - A_2^2 - A_3^2
$$

$$
- \gamma^2 (1 - \beta^2) A_0^2 - \gamma^2 (1 - \beta^2) A_1^2 - A_2^2 - A_3^2
$$

$$
= (A_0)^2 - (A_1)^2 - (A_2)^2 - (A_3)^2 \tag{4.8}
$$

shows that the length of a 4-vector is invariant under Lorentz transformation, which is equivalent to a rotation of axes.

4.2.1   Some Properties of 4-Vectors

Linear Combinations If A and B are two 4-vectors, then $C = aA + bB$ is also a 4-vector for a and b any two constants. This is evident from the linearity of Eq. 4.6. We can check this by taking the temporal component of C.

$$
\begin{aligned}
C_0 = aA_0 + bB_0 &= a\gamma(A_0' + \beta A_1') + b\gamma(B_0' + \beta B_1') \\
&= \gamma(aA_0' + bB_0') + \beta\gamma(aA_1' + bB_1') \\
&= \gamma\left(C_0' + \beta C_1'\right).
\end{aligned}
\tag{4.9}
$$

Similarly, the transformation of other components can be shown.

Inner-Product Invariance The inner product of any two arbitrary 4-vectors A and B is:

$$
\begin{aligned}
A \cdot B &= A_0 B_0 - A_1 B_1 - A_2 B_2 - A_3 B_3 \\
&\equiv A_0 B_0 - \mathbf{A} \cdot \mathbf{B}
\end{aligned}
\tag{4.10}
$$

One can show that $A \cdot B$ is invariant, i.e., $A \cdot B = A' \cdot B'$.
Using Eq. 4.6:

$$
\begin{aligned}
A \cdot B &= A_0 B_0 - A_1 B_1 - A_2 B_2 - A_3 B_3 \\
&= \left[\gamma(A_0' + \beta A_1')\right]\left[(\gamma(B_0' + \beta B_1'))\right] \\
&\quad - \left[\gamma(A_1' + \beta A_0')\right]\left[\gamma(B_1' + \beta B_0')\right] - A_2' B_2' - A_3' B_3' \\
&= \gamma^2\left[A_0' B_0' + \beta(A_0' B_1' + A_1' B_0') + v^2 A_1' B_1'\right] \\
&\quad - \gamma^2\left[A_1' B_1' + \beta(A_1' B_0' + A_0' B_1') + v^2 A_0' B_0'\right] - A_2' B_2' - A_3' B_3' \\
&= A_0' B_0' \cdot \gamma^2(1 - \beta^2) - A_1' B_1' \cdot \gamma^2(1 - \beta^2) - A_2' B_2' - A_3' B_3' \\
&= A_0' B_0' - A_1' B_1' - A_2' B_2' - A_3' B_3' \\
&= A' \cdot B'.
\end{aligned}
\tag{4.11}
$$

Like in 3-dimensional space $\mathbf{A} \cdot \mathbf{B}$ is invariant under rotation, this scalar product of two 4-vectors is invariant in the 4-dimensional Minkowski space.

Further, it can be shown that the *norm*, the square root of an inner product of a 4-vector with itself, is invariant. Now, the inner product of a 4-vector with itself is:

$$|A|^2 \equiv A \cdot A$$
$$= A_0 A_0 - A_1 A_1 - A_2 A_2 - A_3 A_3 \tag{4.12}$$
$$= A_0^2 - |\mathbf{A}|^2.$$

Since the scalar product is invariant, the norm $\sqrt{A \cdot A}$ is also an invariant quantity. This is analogous to $\sqrt{\mathbf{A} \cdot \mathbf{A}} \equiv |\mathbf{A}|$ in 3-dimensional space for rotational invariance. The invariance of $E^2 - p^2 = m_0^2$ and $dt^2 - dx^2 = ds^2$ are the examples of 4-vector norm invariance.

4.2.2 Lorentz Transformation in Four Vector Form

In a similar line as discussed earlier, following Fig. 1.2, consider a Lorentz boost in positive X-direction. The coordinates (t, x, y, z) are in the frame S and the coordinates (t', x', y', z') are in the S' frame, which is moving with velocity $\mathbf{v} = v\hat{i}$. The origins of S and S' are assumed to coincide at $t = t' = 0$. In the four-dimensional Minkowski space, let us assume $(t \equiv x^0, x \equiv x^1, y \equiv x^2, z \equiv x^3)$, namely, the zeroth component of the space-time 4-vector is the temporal component, and the other three components (1, 2, and 3) correspond to the spatial components. Then the two coordinates are related by the following equations:

$$t' = \frac{t - \left(v/c^2\right) x}{\sqrt{1 - (v/c)^2}} \Rightarrow x^{0'} = \gamma \left(x^0 - \beta x^1\right), \tag{4.13}$$

$$x' = \frac{x - vt}{\sqrt{1 - (v/c)^2}} \Rightarrow x^{1'} = \gamma \left(x^1 - \beta x^0\right), \tag{4.14}$$

$$x^{2'} = x^2, \tag{4.15}$$

$$x^{3'} = x^3 \tag{4.16}$$

where, $\beta = v/c$, $\gamma = \frac{1}{\sqrt{1-\beta^2}}$. The above equations can be written in matrix form as:

$$\begin{bmatrix} x^{0'} \\ x^{1'} \\ x^{2'} \\ x^{3'} \end{bmatrix} = \begin{bmatrix} \gamma & -\beta\gamma & 0 & 0 \\ -\beta\gamma & \gamma & 0 & 0 \\ 0 & 0 & 1 & 0 \\ 0 & 0 & 0 & 1 \end{bmatrix} \begin{bmatrix} x^0 \\ x^1 \\ x^2 \\ x^3 \end{bmatrix} \tag{4.17}$$

$$\Rightarrow \begin{bmatrix} x^{0'} \\ x^{1'} \\ x^{2'} \\ x^{3'} \end{bmatrix} = \begin{bmatrix} \Lambda^0_0 & \Lambda^0_1 & \Lambda^0_2 & \Lambda^0_3 \\ \Lambda^1_0 & \Lambda^1_1 & \Lambda^1_2 & \Lambda^1_3 \\ \Lambda^2_0 & \Lambda^2_1 & \Lambda^2_2 & \Lambda^2_3 \\ \Lambda^3_0 & \Lambda^3_1 & \Lambda^3_2 & \Lambda^3_3 \end{bmatrix} \begin{bmatrix} x^0 \\ x^1 \\ x^2 \\ x^3 \end{bmatrix} \tag{4.18}$$

where, $\Lambda^0_0 = \Lambda^1_1 = \gamma$, $\Lambda^0_1 = \Lambda^1_0 = -\gamma\beta$, $\Lambda^2_2 = \Lambda^3_3 = 1$, and all other Λ's are zero. In compact form, Eq. 4.18 can be written as:

$$\boxed{x^{\mu'} = \sum_{\nu=0}^{3} \Lambda^\mu_\nu x^\nu} \quad (\mu = 0, 1, 2, 3) \tag{4.19}$$

Using Einstein's summation convention (repeated Greek indices are to be summed), the above equation can be written as

$$\boxed{x^{\mu'} = \Lambda^\mu_\nu x^\nu} \tag{4.20}$$

In Λ^μ_ν, the superscript μ denotes the row and the subscript ν denotes columns. Λ^μ_ν are the coefficients of a matrix Λ, such that:

$$\Lambda = \begin{bmatrix} \Lambda^0_0 & \Lambda^0_1 & \Lambda^0_2 & \Lambda^0_3 \\ \Lambda^1_0 & \Lambda^1_1 & \Lambda^1_2 & \Lambda^1_3 \\ \Lambda^2_0 & \Lambda^2_1 & \Lambda^2_2 & \Lambda^2_3 \\ \Lambda^3_0 & \Lambda^3_1 & \Lambda^3_2 & \Lambda^3_3 \end{bmatrix} = \begin{bmatrix} \gamma & -\beta\gamma & 0 & 0 \\ -\beta\gamma & \gamma & 0 & 0 \\ 0 & 0 & 1 & 0 \\ 0 & 0 & 0 & 1 \end{bmatrix} \tag{4.21}$$

By taking $\gamma = \sqrt{1 - \beta^2} = \cosh\omega$, so that $\sinh\omega = \gamma\beta$, the above matrix can be written as:

$$\Lambda = \begin{bmatrix} \cosh\omega & -\sinh\omega & 0 & 0 \\ -\sinh\omega & \cosh\omega & 0 & 0 \\ 0 & 0 & 1 & 0 \\ 0 & 0 & 0 & 1 \end{bmatrix} \tag{4.22}$$

In analogy with the 3-dimensional rotation matrix, it is easy to identify Eq. 4.22 as the rotation matrix in the Minkowski space. Comparing with Eq. 4.21, one observes that the Lorentz transformation is nothing but a rotation in the 4-dimensional Minkowski space.

Using 4-vector inner product, one can easily show that, like the invariance of the 3-vector under rotation, $i.e.,$ $\mathbf{X}'^2 = \mathbf{X}^2$, the quantity, $(x^0)^2 - (\mathbf{X})^2$ is invariant under the Lorentz transformation.

4.2.3 The Position-Time 4-Vector

The Lorentz invariant quantity, $(x^0)^2 - (\mathbf{X})^2$ can be written in a more compact form as follows.

$$x^\mu = (x^0, x^1, x^2, x^3) = (x^0, \mathbf{X}) \tag{4.23}$$

with $\mu = 0, 1, 2, 3$ and $x^0 = t, x^1 = x, x^2 = y, x^3 = z$.

$$x_\mu = (x_0, x_1, x_2, x_3) = (x_0, -\mathbf{X}) \tag{4.24}$$

In this convention, $x_0 = x^0$, $x_1 = -x^1$, $x_2 = -x^2$, and $x_3 = -x^3$, so that a *"covariant 4-vector"*, x_μ (index down):

$$x_\mu = g_{\mu\nu} x^\nu, \tag{4.25}$$

where the *spacetime metric*

$$g_{\mu\nu} = \begin{bmatrix} 1 & 0 & 0 & 0 \\ 0 & -1 & 0 & 0 \\ 0 & 0 & -1 & 0 \\ 0 & 0 & 0 & -1 \end{bmatrix}. \tag{4.26}$$

Here, $g_{\mu\nu}$ is the *"flat spacetime"* metric in the Minkowski space.

x^μ (index up) is called *"contravariant 4-vector"*.

$g^{\mu\nu}$ are the elements in g^{-1}. Since $g^{-1} = g$, $\boxed{g^{\mu\nu} = g_{\mu\nu}}$.

Using the position-time 4-vector, x^μ, one can show that the 4-dimensional length element

$$\begin{aligned} I &= x_\mu x^\mu = x^\mu x_\mu \\ I &\equiv (x^0)^2 - (x^1)^2 - (x^2)^2 - (x^3)^2 \\ &= (x^{0'})^2 - (x^{1'})^2 - (x^{2'})^2 - (x^{3'})^2, \end{aligned} \tag{4.27}$$

which is Lorentz Invariant (LI). A quantity having the same value in all inertial frames is called an *"invariant"*. This is like $r^2 = x^2 + y^2 + z^2$ being invariant under spatial rotation.

Following the above discussion, to each contravariant 4-vector a^μ, a covariant 4-vector could be assigned and vice versa.

$$a^\mu = g^{\mu\nu} a_\nu \tag{4.28}$$

$$a_\mu = g_{\mu\nu} a^\nu \tag{1.29}$$

Given any two 4-vectors, a^μ and b^μ,

$$a^\mu b_\mu = a_\mu b^\mu = a^0 b^0 - a^1 b^1 - a^2 b^2 - a^3 b^3 \qquad (4.30)$$

is a Lorentz invariant quantity. The above operation is also called "4-vector scalar product".

$$a.b \equiv a_\mu b^\mu$$
$$= a^0 b^0 - \mathbf{a.b} \qquad (4.31)$$

$$a^2 \equiv a.a = (a^0)^2 - \mathbf{a}^2 \qquad (4.32)$$

The 1st term is called "*temporal*" and the 2nd is called "*spatial*" component.

- If $a^2 > 0$: a^μ is called *time-like*. Events are in the forward light-cone. They appear later than the origin, O. Events in the backward light-cone appear earlier to O. Only events in the backward light-cone can influence O. And O can have an influence only on the events in the forward cone.
- If $a^2 < 0$: a^μ is called *space-like*. Events are called space-like events, and there is no interaction with O. This is related to "*causality*". This required $v > c$.
- If $a^2 = 0$: a^μ is called *light-like*. Connects all those events with the origin, which can be reached by a light signal.

This is shown pictorially in Fig. 4.2, taking spacetime 4-vector, which is called the lightcone diagram or the Minkowski spacetime diagram. A *worldline* is the curve connecting all spacetime points that describe the history of an object.

4.2.4   The Displacement 4-Vector

If an event A occurs at $\left(x_A^0, x_A^1, x_A^2, x_A^3\right)$ and another event B occurs at $\left(x_B^0, x_B^1, x_B^2, x_B^3\right)$, then the displacement 4-vector is:

$$dx^\mu = x_A^\mu - x_B^\mu. \qquad (4.33)$$

4.2.5   Spacetime Interval Between Two Events

The 4-vector scalar product of dx^μ with itself gives the spacetime interval between two events.

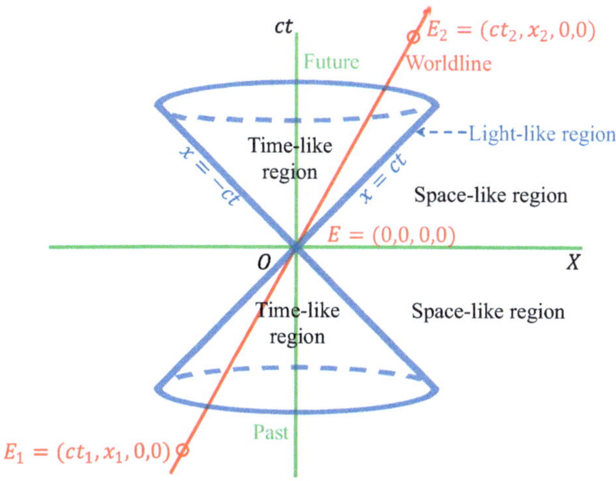

Fig. 4.2 A schematic of a light-cone diagram or Minkowski spacetime diagram. A Minkowski spacetime light-cone diagram shows the different causal regions corresponding to an event $E = (0, 0, 0, 0)$ at the origin. The time-like regions (inside the light cones) are regions where there can be a causal relationship between the event at the origin and an event within the time-like regions (where $v < c$). The spacelike regions (outside the light cones) are regions where there cannot be a causal relationship between the event at the origin and any event in the spacelike region (because $v > c$). A worldline is shown for an object with mass moving at a constant speed $v < c$. Worldlines of objects with mass can move only through time-like regions. Events in light-like regions (the two cones themselves) can also be causally connected by light-speed photon transfer

$$
\begin{aligned}
(ds)^2 &= dx_\mu dx^\mu \\
&= (dx^0)^2 - \left[(dx^1)^2 + (dx^2)^2 + (dx^3)^2 \right] \\
&= (dt)^2 - d^2,
\end{aligned}
\tag{4.34}
$$

where dt is the time interval between the two events having a spatial separation of d.

- For $(dt)^2 > d^2, \Rightarrow (ds)^2 > 0$, the intervals are real and are time-like intervals.
- For $(dt)^2 < d^2, \Rightarrow (ds)^2 < 0$, the intervals are imaginary and are space-like intervals. For a reference frame, where the two events are simultaneous, $dt = 0$ and hence, $(ds)^2 = -d^2 \Rightarrow ds = i\,d$.
- For $(dt)^2 = d^2, \Rightarrow (ds)^2 = 0$, the intervals are called zero intervals. This occurs between events connected by a light signal. These are light-like intervals.

4.2.6 The Proper Time Interval ($d\tau$)

Let us consider an object which is at rest in an inertial frame S' that moves with a uniform velocity $\mathbf{v}$ along the common $X - X' - axis$, relative to the stationary inertial frame S. Both S and S' frames had a common origin at $t = t' = 0$. Suppose two events occur in S' at the same spatial point ($dx^{1'} = dx^{2'} = dx^{3'} = 0$) but with a time interval dt'.

Now, keeping the speed of light in the calculation, for a better handle,

$$ds^2 = c^2 dt'^2 - \left[\left(dx^{1'}\right)^2 + \left(dx^{2'}\right)^2 + \left(dx^{3'}\right)^2 \right]$$

$$= c^2 dt'^2 \tag{4.35}$$

$$\Rightarrow dt' = \frac{ds}{c}.$$

As discussed earlier, the time interval dt' measured in the S' frame, where the object is stationary, is called the proper time interval and is denoted by $d\tau$. Hence,

$$d\tau = \frac{ds}{c}. \tag{4.36}$$

Both ds and c are Lorentz invariant quantities. Hence, $d\tau$ is also a Lorentz invariant quantity. dt is the time interval measured by an observer in the frame S. Now, let us proceed to show the relationship between $d\tau$ and dt from intervals.

$$ds^2 = (cdt)^2 - (dx^1)^2 - (dx^2)^2 - (dx^3)^2$$

$$= (cdt)^2 \left[1 - \frac{(dx^1)^2}{c^2(dt)^2} - \frac{(dx^2)^2}{c^2(dt)^2} - \frac{(dx^3)^2}{c^2(dt)^2} \right]$$

$$= (cdt)^2 \left(1 - \frac{v^2}{c^2} \right)$$

$$= (cdt)^2 \left(1 - \beta^2 \right)$$

$$\Rightarrow \frac{ds^2}{c^2} = (dt)^2 \left(1 - \beta^2 \right) \text{ (by using Eq. 4.36)}$$

$$\Rightarrow d\tau^2 = (dt)^2 \left(1 - \beta^2 \right)$$

$$\Rightarrow d\tau = dt\sqrt{\left(1 - \beta^2 \right)}$$

Proper time, $\boxed{d\tau \equiv dt\sqrt{1 - \beta^2}}$ is a Lorentz invariant scalar, which we also saw earlier using Lorentz transformation equations.

4.2.7 The Velocity 4-Vector

The velocity of a particle is given by

$$\mathbf{v} = \frac{d\mathbf{x}}{dt} \tag{4.37}$$

where $d\mathbf{x}$ is the distance travelled in the laboratory frame (frame-S) and dt is the time measured in the same frame. The proper velocity of the particle is given by

$$\boldsymbol{\eta} = \frac{d\mathbf{x}}{d\tau} \tag{4.38}$$

where $d\mathbf{x}$ is the distance travelled in the laboratory frame and $d\tau$ is the proper time (time measured in the rest frame of the particle. For simplicity, we consider the particle frame to be the S'-frame). Now

$$\boldsymbol{\eta} = \frac{d\mathbf{x}}{d\tau} = \frac{d\mathbf{x}}{dt}\frac{dt}{d\tau}$$
$$= \mathbf{v}\gamma$$
$$\Rightarrow \boxed{\boldsymbol{\eta} = \gamma\mathbf{v}} \tag{4.39}$$

It is easy to work with the proper velocity, $\boldsymbol{\eta}$, as only $d\mathbf{x}$ transforms under the Lorentz transformation, while $d\tau$ is an invariant quantity. Furthermore,

$$\eta^{\mu} = \frac{dx^{\mu}}{d\tau} \tag{4.40}$$

so

$$\eta^{0} = \frac{dx^{0}}{d\tau} = \frac{d(ct)}{\frac{1}{\gamma}dt} = \gamma c \tag{4.41}$$

Hence

$$\boxed{\eta^{\mu} = \gamma\left(c,\ v_x,\ v_y,\ v_z\right) = \gamma\left(c,\ \mathbf{v}\right)} \tag{4.42}$$

This is called the *proper velocity 4-vector*. Remember that the spatial component brings up the negative sign for the covariant tensor. Now

$$\eta^{\mu}\eta_{\mu} = \gamma^2\left(c^2 - v_x^2 - v_y^2 - v_z^2\right)$$
$$= \gamma^2 c^2\left(1 - \frac{v^2}{c^2}\right)$$

$$= \gamma^2 c^2 \left(1 - \beta^2\right) = c^2$$

$$\Rightarrow \boxed{\eta^2 = c^2},$$
(4.43)

which is Lorentz Invariant. This also proves that the 4-vector scalar product is L.I.

4.2.8 The Acceleration 4-Vector

The acceleration 4-vector is obtained by taking the derivative of the velocity 4-vector with respect to proper time, τ, *i.e.*,

$$\begin{aligned}
A^\mu &= \frac{d\eta^\mu}{d\tau} \\
&= \frac{d}{d\tau}(\gamma c, \gamma \mathbf{v})
\end{aligned}$$
(4.44)

The temporal component,

$$\begin{aligned}
A^0 &= \frac{d\eta^0}{d\tau} \\
&= \frac{d(\gamma c)}{d\tau} = \frac{d(\gamma c)}{dt/\gamma} = c\gamma\dot{\gamma},
\end{aligned}$$
(4.45)

where $\dot{\gamma} = \frac{d\gamma}{dt}$.

The spatial component

$$\begin{aligned}
A^i &= \frac{d\eta^i}{d\tau} \\
&= \frac{d(\gamma v^i)}{d\tau} = \frac{d(\gamma v^i)}{dt/\gamma} = \gamma^2 \mathbf{a} + \gamma\dot{\gamma}v^i,
\end{aligned}$$
(4.46)

where the 3-acceleration, $\mathbf{a} = \frac{d\mathbf{v}}{dt}$ and $v^i = \mathbf{v}$ is the 3-vector.

Now,

$$\dot{\gamma} = \frac{d}{dt}\left[\frac{1}{\sqrt{1 - v^2/c^2}}\right] = \frac{\mathbf{a}\cdot\mathbf{v}}{c^2}\gamma^3$$
(4.47)

Using Eq. 4.47 in the above two equations, we get:

$$
\begin{aligned}
A^\mu &= \left(c\gamma\dot{\gamma}, \gamma^2 \mathbf{a} + \gamma\dot{\gamma}\mathbf{v} \right) \\
&= \left(c\gamma \frac{\mathbf{a}\cdot\mathbf{v}}{c^2}\gamma^3, \gamma^2 \mathbf{a} + \gamma \frac{\mathbf{a}\cdot\mathbf{v}}{c^2}\gamma^3 \mathbf{v} \right) \\
&= \left(\gamma^4 \frac{\mathbf{a}\cdot\mathbf{v}}{c}, \gamma^2 \mathbf{a} + \gamma^4 \frac{\mathbf{a}\cdot\mathbf{v}}{c^2}\mathbf{v} \right)
\end{aligned}
\tag{4.48}
$$

Example 4.2 Show that the 4-velocity and the 4-acceleration are mutually orthogonal.

Solution 4.2 The acceleration 4-vector is:

$$
\begin{aligned}
A^\mu &= \frac{d\eta^\mu}{d\tau} \\
&= \frac{d^2 x^\mu}{d\tau^2}
\end{aligned}
\tag{4.49}
$$

Differentiating $\eta^\mu \eta_\mu = c^2$ with respect to τ, we get:

$$
\begin{aligned}
\frac{d\eta^\mu}{d\tau}\eta_\mu + \eta^\mu \frac{d\eta_\mu}{d\tau} &= 0 \\
\rightarrow A^\mu \eta_\mu + \eta^\mu A_\mu &= 0 \\
\Rightarrow A^\mu \eta_\mu + A^\mu \eta_\mu &= 0 \\
\rightarrow 2 A^\mu \eta_\mu &= 0 \\
\Rightarrow A \cdot \eta &= 0
\end{aligned}
\tag{4.50}
$$

This shows that the 4-velocity and the 4-acceleration are mutually orthogonal.

4.2.9 Energy-Momentum Four-Vector

We know *momentum* $=$ *rest mass* $\times$ *velocity*. Velocity can be either "ordinary velocity" or "proper velocity". Classically, both are equal (non-relativistic limit). If $\mathbf{p} = m_0 \mathbf{v}$, the conservation of momentum is inconsistent with the principle of relativity. In relativity, momentum is the product of mass and proper velocity.

$$
\mathbf{p} \equiv m_0 \eta
\tag{4.51}
$$

Momentum 4-vector or four momentum is given by:

$$
p^\mu = m_0 \eta^\mu
\tag{4.52}
$$

The spatial component of p^μ constitutes the (relativistic) momentum 4-vector:

$$\mathbf{p} = \gamma m_0 \mathbf{v} = \frac{m_0 \mathbf{v}}{\sqrt{1 - v^2/c^2}} \tag{4.53}$$

$$\boxed{p^0 = \gamma m_0 c} \tag{4.54}$$

Relativistic energy, E:

$$E \equiv \gamma m_0 c^2 = \frac{m_0 c^2}{\sqrt{1 - v^2/c^2}} \tag{4.55}$$

Hence,

$$p^0 = \frac{E}{c} \tag{4.56}$$

and the energy-momentum 4-vector:

$$\boxed{p^\mu = \left(\frac{E}{c},\ p_x,\ p_y,\ p_z\right) = \left(\frac{E}{c}, \mathbf{p}\right) = (\gamma m_0 c, \gamma m_0 \mathbf{v})} \tag{4.57}$$

Now

$$p^\mu p_\mu = \frac{E^2}{c^2} - \mathbf{p}^2 = m_0^2 c^2$$

$$= \left(m_0 \eta^\mu\right)\left(m_0 \eta_\mu\right)$$

$$= m_0^2 \left(\eta^\mu \eta_\mu\right)$$

$$= m_0^2 c^2$$

$$\Rightarrow \boxed{p^\mu p_\mu \equiv p^2 = m_0^2}\ (Natural\ Units). \tag{4.58}$$

The right-hand side is a Lorentz invariant quantity and hence, as expected from a 4-vector scalar product the left-hand side is also Lorentz invariant.

$$\Rightarrow E^2 = \mathbf{p}^2 c^2 + m_0^2 c^4$$

In natural units,

$$\boxed{E^2 = \mathbf{p}^2 + m_0^2} \tag{4.59}$$

As energy and momentum constitute p^μ, it is called *"energy-momentum 4-vector"*.

- $p^2 = m_0^2 > 0$: Particles with real and positive rest mass. These particles always travel with a velocity less than c, and are represented by a hyperbola in the forward light-cone.
- $p^2 = m_0^2 = 0$: Zero rest mass particles like photon, graviton, neutrino, *etc.* These particles travel with the speed of light and are represented by the degenerate limit of the hyperbola in the forward light-cone.
- $p^2 = m_0^2 < 0$: For this class of particles, the rest mass is imaginary, like tachyon (theorized) or virtual particles. They travel at a speed higher than the speed of light.
- $p^\mu = 0$: Vacuum

Remember that the relativistic equations $\mathbf{p} = \gamma m_0 \mathbf{v}$ and $E = \gamma m_0$ do not hold good for massless particles, and $m_0 = 0$ is allowed only if the particle travels with the speed of light. For massless particles,

$$v = c \text{ and } E = |\mathbf{p}|c.$$

Example 4.3 The change in momentum and the change in energy.

Solution 4.3 The problem is to find the relation between the fractional change in the momentum of a particle with the fractional change in its energy. We know the relativistic formula for the energy of a particle with rest mass, m_0, is given by

$$E = \sqrt{p^2 + m_0^2} \text{ (in natural unit)}$$

$$\implies \frac{dE}{dp} = \frac{d}{dp}\left(\left[p^2 + m_0^2\right]^{1/2}\right)$$

$$= \frac{1}{2}\left[p^2 + m_0^2\right]^{-1/2} \times 2p$$

$$= \frac{p}{E} = \frac{E}{p}\left(\frac{p}{E}\right)^2$$

$$= \frac{E}{p}\left(\frac{\gamma m_0 \beta}{\gamma m_0}\right)^2 = \frac{E}{p}\beta^2$$

$$\implies \frac{dp}{p} = \frac{1}{\beta^2}\frac{dE}{E} \tag{4.60}$$

4.2.10 The Force 4-Vector

The force 4-vector, K^μ, is defined as the derivative of 4-momentum with respect to τ.

$$\boxed{K^\mu = \frac{dp^\mu}{d\tau}} \tag{4.61}$$

The force 4-vector is also called the Minkowski force. Let us now write the temporal and spatial components of the force 4-vector.

We know, $p^\mu = m_0 \eta^\mu$ and $\eta^\mu = \frac{dx^\mu}{d\tau}$. Hence,

$$\begin{aligned}
K^\mu &= m_0 \frac{d\eta^\mu}{d\tau} = m_0 A^\mu \\
&= m_0 \frac{d^2 x^\mu}{d\tau^2},
\end{aligned} \tag{4.62}$$

where, A^μ is the 4-acceleration. The force 4-vector can also be defined as the rest mass times the acceleration 4-vector. Further,

$$K^\mu = \frac{dt}{d\tau} \frac{dp^\mu}{dt} = \gamma \frac{dp^\mu}{dt}. \tag{4.63}$$

Now the temporal component of the four force is:

$$K^0 = \gamma \frac{dp^0}{dt} = \frac{\gamma}{c} \frac{dE}{dt}. \tag{4.64}$$

K^0 is $\frac{\gamma}{c}$ times the rate of change of energy, i.e., the power delivered to the particle. As

$$\frac{dE}{dt} = \mathbf{F} \cdot \mathbf{v}, \tag{4.65}$$

$$K^0 = \frac{\gamma}{c} \mathbf{F} \cdot \mathbf{v}. \tag{4.66}$$

Now the spatial component of the four force is:

$$K^i = \gamma \frac{dp^i}{dt} = \gamma F^i. \quad \text{(with } i = 1,2,3\text{)}, \tag{4.67}$$

where $F^i = \frac{dp^i}{dt}$ is the ith component of the three vector force, $\mathbf{F}$. Hence, the spatial component of the four force is $\mathbf{K} = \gamma \mathbf{F}$.

Combining both the components, the four force is:

$$K^\mu = \left(\frac{\gamma}{c}\mathbf{F}\cdot\mathbf{v}, \gamma\mathbf{F}\right)$$

$$= m_0 A^\mu = m_0 \left(\gamma^4\frac{\mathbf{a}\cdot\mathbf{v}}{c}, \gamma^2\mathbf{a} + \gamma^4\frac{\mathbf{a}\cdot\mathbf{v}}{c^2}\mathbf{v}\right) \tag{4.68}$$

Here we have used Eq. 4.48.

4.2.11 Decay of an Unstable Particle: Finding the Energies and Momenta of Daughter Particles in a Two-Body Decay Using Conservation of 4-Momentum

Consider the decay (in natural units $c = 1$) of an unstable particle of rest mass M and 4–momentum P^μ to two daughter particles having masses m_1, m_2 and 4–momenta p_1^μ, p_2^μ.

$$A(P) \;\rightarrow\; 1(p_1) + 2(p_2),$$

Case-1: In the *rest frame of the parent A*

As the mother particle is at rest, the daughters will be emitted back-to-back, conserving energy-momentum. Hence, the 4-momenta of the associated particles are:

$$P^\mu = (M, \mathbf{0}), \qquad p_1^\mu = (E_1, \mathbf{p}), \qquad p_2^\mu = (E_2, -\mathbf{p}).$$

The conservation of 4-momentum is:

$$P^\mu = p_1^\mu + p_2^\mu$$

$$\Rightarrow p_2^\mu = P^\mu - p_1^\mu$$

$$\Rightarrow p_2^\mu p_{2\mu} = (P^\mu - p_1^\mu)(P_\mu - p_{1\mu})$$

$$\Rightarrow m_2^2 = M^2 + m_1^2 - 2(EE_1 - \mathbf{P}.\mathbf{p_1})$$

$$\Rightarrow m_2^2 = M^2 + m_1^2 - 2(ME_1 - |\mathbf{P}||\mathbf{p_1}|\cos\theta) \quad (\because \mathbf{P} = 0) \tag{4.69}$$

$$\Rightarrow m_2^2 = M^2 + m_1^2 - 2ME_1$$

$$\Rightarrow E_1 = \frac{M^2 + m_1^2 - m_2^2}{2M}$$

[Recall, the square of a 4-momentum is the square of the rest mass of the particle.]

Similarly, one can show that:

$$E_2 = \frac{M^2 + m_2^2 - m_1^2}{2M} \tag{4.70}$$

Hence, the energies of the daughter particles, when the mother particle decays at rest, are:

$$\boxed{E_1 = \frac{M^2 + m_1^2 - m_2^2}{2M}, \qquad E_2 = \frac{M^2 + m_2^2 - m_1^2}{2M}}$$

This satisfies $E_1 + E_2 = M$, as expected from the temporal part of the 4-momentum conservation, i.e., conservation of energy. Now, let us find the magnitude of the daughter's momentum. For this, we use the on-shell formula,

$$E_1^2 = p_1^2 + m_1^2$$

$$\Rightarrow p_1^2 = E_1^2 - m_1^2$$

$$= \left[\frac{M^2 + m_1^2 - m_2^2}{2M}\right]^2 - m_1^2 \tag{4.71}$$

$$\Rightarrow |\mathbf{p_1}| = |\mathbf{p_2}| \equiv p = \frac{\left(M^4 + m_1^4 + m_2^4 - 2M^2m_1^2 - 2M^2m_2^2 - 2m_1^2m_2^2\right)^{1/2}}{2M}$$

Using the Källén function $\lambda(a, b, c) = a^2 + b^2 + c^2 - 2ab - 2bc - 2ca$,

$$p^2 = \frac{\lambda(M^2, m_1^2, m_2^2)}{4M^2}.$$

Now,

$$p = \frac{\left(M^4 + m_1^4 + m_2^4 - 2M^2m_1^2 - 2M^2m_2^2 - 2m_1^2m_2^2\right)^{1/2}}{2M}$$

$$= \frac{\left(M^4 + m_1^4 + m_2^4 - 2M^2m_1^2 - 2M^2m_2^2 + 2m_1^2m_2^2 - 4m_1^2m_2^2\right)^{1/2}}{2M}$$

$$= \frac{\left[\left(M^2 - m_1^2 - m_2^2\right)^2 - 4m_1^2m_2^2\right]^{1/2}}{2M} \tag{4.72}$$

$$= \frac{\left\{\left[\left(M^2 - m_1^2 - m_2^2\right) - 2m_1m_2\right]\left[\left(M^2 - m_1^2 - m_2^2\right) + 2m_1m_2\right]\right\}^{1/2}}{2M}$$

$$= \frac{\left\{\left[M^2 - \left(m_1^2 + m_2^2 + 2m_1m_2\right)\right]\left[M^2 - \left(m_1^2 + m_2^2 - 2m_1m_2\right)\right]\right\}^{1/2}}{2M}$$

$$= \frac{\left\{\left[M^2 - (m_1 + m_2)^2\right]\left[M^2 - (m_1 - m_2)^2\right]\right\}^{1/2}}{2M}$$

A more simplified formula for the magnitude of the daughter 3-momentum (which is common, back–to–back for both the particles) is

$$p = \frac{\sqrt{\left[M^2 - (m_1 + m_2)^2\right]\left[M^2 - (m_1 - m_2)^2\right]}}{2M}$$

As special cases,

- If $m_1 = m_2 = m$ then $E_1 = E_2 = M/2$ and $p = \frac{1}{2}\sqrt{M^2 - 4m^2}$.
- If $m_2 = 0$ (massless particle), $E_2 = \dfrac{M^2 - m_1^2}{2M}$, $E_1 = \dfrac{M^2 + m_1^2}{2M}$, and $p = E_2$.

Case-2: Energies in a lab frame where *the parent is moving*

Let the parent have lab 4–momentum $P^\mu_{\text{lab}} = (E_A, \mathbf{P})$. The velocity of the parent (lab frame) is

$$\mathbf{v} = \frac{\mathbf{P}}{E_A}, \qquad \gamma = \frac{E_A}{M}.$$

Boosting a daughter from the parent rest frame to the lab (boost along the direction of $\mathbf{P}$) gives the lab energy of daughter 1. For this, we use the inverse Lorentz transformation for energy:

$$E_1 = \gamma\left(E_1^* + v\, p^* \cos\theta^*\right)$$

where θ^* is the angle between $\mathbf{p}^*$ (in the parent rest frame) and the boost direction; $v = |\mathbf{v}|$. Similarly for daughter 2 (since $\mathbf{p}_2^* = -\mathbf{p}_1^*$):

$$E_2 = \gamma\left(E_2^* - v\, p^* \cos\theta^*\right).$$

Using $\gamma = E_A/M$ and $|\mathbf{P}| = \gamma M v$ one may also write

$$E_1 = \frac{E_A}{M} E_1^* + \frac{|\mathbf{P}|}{M} p^* \cos\theta^*.$$

If the decay is isotropic in the parent rest frame, $\cos\theta^*$ is uniformly distributed in $[-1, 1]$ and the lab energy E_1 ranges between

$$E_1^{\max/\min} = \gamma\left(E_1^* \pm v\, p^*\right).$$

Here, a star($*$) denotes quantities evaluated in the parent rest frame, which essentially refer to the quantities estimated in our Case-1. In Case-1, we didn't keep the ($*$) knowingly to avoid confusion and to make it an independent problem by itself.

It should be noted here that Case-2 can be dealt with independently using the 4-momentum conservation, without using a Lorentz boost on Case-1, which we leave to the readers to try out and have fun with 4-vectors.

Example 4.4 Production of π^0 mesons: Estimate the minimum kinetic energy of the incident proton that collides with a target proton at rest to produce a π^0 meson.

Solution 4.4 This is a kinematic problem in high-energy physics, where π^0 mesons are produced in the laboratory by colliding energetic protons on proton targets at rest. The reaction is:

$$p + p \rightarrow p + p + \pi^0.$$

The 4-momenta of the involved particles are:

Projectile proton: $p_a^\mu = (E_a, \mathbf{p_a})$, for the target proton which is at rest, $p_b^\mu = (m_p, \mathbf{0})$. For the final state involving two protons and a π^0 meson, $p_f^\mu = (2m_p + m_{\pi^0}, \mathbf{0})$. Note that, as we are estimating the minimum kinetic energy of the projectile proton, the final state particles are just produced with zero kinetic energy, which will be later termed as threshold production. Let us hold on with the discussion of *"threshold energy"* till the subsequent sections.

Now, the 4-momenta of the system before the interaction is: $p_{before}^\mu = (E_p + m_p, \mathbf{p})$ and the 4-momenta of the system after the interaction is: $p_{after}^\mu = (2m_p + m_{\pi^0}, \mathbf{0})$.

$\left(p_{total}^\mu\right)^2$ before and after the interaction is a Lorentz invariant quantity. Hence,

$$p_{before}^\mu \cdot p_{before\,\mu} = p_{after}^\mu \cdot p_{after\,\mu}$$

$$\Rightarrow (E_p + m_p)^2 - \mathbf{p}^2 = (2m_p + m_{\pi^0})^2$$

$$\Rightarrow E_p^2 + m_p^2 + 2E_p m_p - p^2 = (2m_p + m_{\pi^0})^2$$

$$\Rightarrow (E_p^2 - p^2) + m_p^2 + 2E_p m_p = (2m_p + m_{\pi^0})^2$$

$$\Rightarrow m_p^2 + m_p^2 + 2E_p m_p = (2m_p + m_{\pi^0})^2 \tag{4.73}$$

$$\Rightarrow 2m_p^2 + 2E_p m_p = (2m_p + m_{\pi^0})^2$$

$$\Rightarrow E_p = \frac{(2m_p + m_{\pi^0})^2 - 2m_p^2}{2m_p}$$

$$= m_p + 2m_{\pi^0} + \frac{m_{\pi^0}^2}{2m_p}$$

Hence, the minimum kinetic energy of the projectile proton is:

$$T = E_p - m_p$$

$$= \left(m_p + 2m_{\pi^0} + \frac{m_{\pi^0}^2}{2m_p} \right) - m_p$$

$$= 2m_{\pi^0} + \frac{m_{\pi^0}^2}{2m_p} \tag{4.74}$$

$$= 2 \times 134.9 + \frac{(134.9)^2}{2 \times 938}$$

$$= 279.5 \text{ MeV}.$$

For the estimation of the minimum kinetic energy, here we have taken the rest mass of π^0 as 134.9 MeV and the proton rest mass 938 MeV.

We shall directly use the estimation of the threshold energy formula to solve this problem in the subsequent sections. However, the method of estimation will be the same, the latter being a more generalized formula, as we shall see soon.

4.2.12 Elegance of 4-Vector Method Over the Traditional One

Now we shall deal with the Compton scattering problem to estimate the wavelength shift of the scattered photon, first using the traditional method of energy and momentum conservation. Then we shall solve the same problem using the discussed 4-vector method to show the elegance of the 4-vector method over the traditional way of solving a kinematic problem.

Compton Scattering: An Application of Special Theory of Relativity
The phenomenon of *elastic scattering of a photon from an electron*, called the Compton effect or Compton scattering, gave astounding evidence for the foundations of quantum mechanics and the particle nature of light, while supporting the special theory of relativity.

A.H. Compton used high-frequency radiation like X-rays and γ-rays on a light element target like graphite, where light-matter interaction gave rise to a reduced energy of the scattered photon from the atomic electron, which recoils during the interaction. The observed change in the frequency or the shift in wavelength of the scattered light is called the Compton effect. The classical theory of an electromagnetic wave in Thomson scattering does not explain this shift in wavelength at low intensities. This effect was very well explained through the quantum theory of radiation with the application of the special theory of relativity. Compton's experiment convincingly explained that light can be treated as a stream of particle-like objects or quanta called photons, whose energy is proportional to the light wave's frequency, i.e., the radiation consists of photons with energy $h\nu$, where ν is the frequency of the radiation and h is Planck's constant. The underlying theory is as follows:

- According to quantum theory, radiation consists of quanta of light called photons having energy $h\nu$.
- These photons travel at the speed of light and obey energy-momentum conservation as they scatter off the atomic electrons, which are loosely bound in the atoms/molecules of light elements.
- As the binding energy of the atoms in light elements is lower than the high-frequency radiation used in Compton scattering, the involved valence electrons behave like free particles.
- When an incident photon with energy $h\nu$ collides with the electron, it transfers some of its energy to this electron. As a matter of fact, the electron gains some kinetic energy to get recoiled, and the photon is scattered off in a different direction with lower energy or lower frequency, or a longer wavelength than the incident photon.

Here, it is assumed that:

- The electron is free and is at rest before the collision.
- Both photon and electron are assumed to be spinless.
- The energy-momentum conservation holds good.

Now, **before the collision:**

The energy of the incident photon $= h\nu$
The momentum of the incident photon $= h\nu/c$
The energy of the stationary electron $= m_e c^2$, m_e being the rest mass of electron.
The momentum of the stationary electron $= 0$.

After the collision:
Let θ and ϕ be respectively the angles made by the scattered photon and the recoiled electron with the direction of the incident photon as shown in Fig. 4.3.

The energy of the scattered photon $= h\nu'$
The momentum of the incident photon $= h\nu'/c$
The energy of the scattered electron $= mc^2$, m is the mass of the electron after the collision. This uses the variation of mass with velocity following the special theory of relativity.
The momentum of the recoiled electron $= mv$, v is the velocity of the electron after the collision.
The total energy of the system before collision $= h\nu + m_e c^2$.
The total energy of the system after collision $= h\nu' + mc^2$.
The total momentum of the system before collision $= h\nu/c + 0 = h\nu/c$.

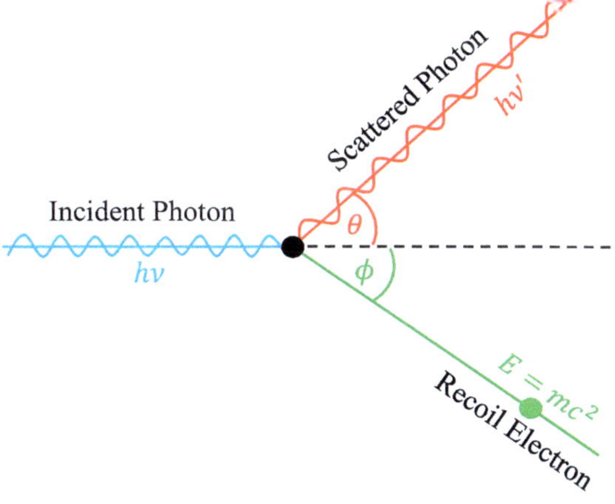

Fig. 4.3 A schematic of elastic scattering of a photon from an electron—The Compton Effect

The total momentum of the system after collision (along the incident photon direction) $= \frac{hv'}{c}\cos\theta + mv\cos\phi$.

Applying conservation of energy:

$$hv + m_e c^2 = hv' + mc^2 \tag{4.75}$$

Applying conservation of momentum along the direction of the incident photon:

$$\frac{hv}{c} = \frac{hv'}{c}\cos\theta + mv\cos\phi \tag{4.76}$$

Now applying conservation of momentum in the direction perpendicular to the incident photon:

$$0 = \frac{hv'}{c}\sin\theta - mv\sin\phi$$
$$\Rightarrow \frac{hv'}{c}\sin\theta = mv\sin\phi \tag{4.77}$$

Rearranging the above two momentum conservation equations:

$$mvc\cos\phi = hv - hv'\cos\theta$$
$$mvc\sin\phi = hv'\sin\theta \tag{4.78}$$

Squaring and adding these two equations, we get:

$$m^2 v^2 c^2 = h^2 \left[v^2 + v'^2 - 2vv' \cos\theta \right]$$
$$\Rightarrow p^2 c^2 = h^2 \left[v^2 + v'^2 - 2vv' \cos\theta \right] \tag{4.79}$$

Here, $p = mv$ is the momentum of the recoiled electron. Equation 4.75 on squaring, gives:

$$hv + m_e c^2 = hv' + mc^2$$
$$\Rightarrow mc^2 = hv - hv' + m_e c^2$$
$$\Rightarrow m^2 c^4 = h^2 (v - v')^2 + m_e{}^2 c^4 + 2m_e c^2 (hv - hv') \tag{4.80}$$
$$\Rightarrow E^2 = h^2 (v^2 + v'^2 - 2vv') + m_e{}^2 c^4 + 2hm_e c^2 (v - v')$$

E is the energy of the recoiled electron.

Subtracting Eq. 4.80 from Eq. 4.79, we get:

$$p^2 c^2 - E^2 = h^2 (v^2 + v'^2 - 2vv' \cos\theta) - h^2 (v^2 + v'^2 - 2vv')$$
$$- 2hm_e c^2 (v - v') - m_e{}^2 c^4$$
$$\Rightarrow -m_e^2 c^4 = h^2 (2vv' - 2vv' \cos\theta) - 2hm_e c^2 (v - v') - m_e^2 c^4$$
$$\Rightarrow 2h^2 vv'(1 - \cos\theta) = 2hm_e c^2 (v - v')$$
$$\Rightarrow \frac{h(1 - \cos\theta)}{m_e c^2} = \frac{(v - v')}{vv'} = \frac{1}{v'} - \frac{1}{v} \tag{4.81}$$
$$\Rightarrow \frac{c}{v'} - \frac{c}{v} = \frac{h}{m_e c}(1 - \cos\theta)$$
$$\Rightarrow (\lambda' - \lambda) \equiv \Delta\lambda = \frac{h}{m_e c}(1 - \cos\theta)$$

Hence, the Compton shift in the wavelength is:

$$\boxed{\Delta\lambda = \frac{h}{m_e c}(1 - \cos\theta)} \tag{4.82}$$

Now, let us solve the same problem using the 4-vector method.

The 4-momenta before the collision are:

$$p_\gamma = \left(\frac{hc}{\lambda}, \frac{hc}{\lambda}, 0, 0 \right), \quad p_e = \left(m_e c^2, 0, 0, 0 \right). \tag{4.83}$$

Note that the energy and momentum for massless particles like photons are the same. Further, the 4-momenta after the collision are:

$$p'_\gamma = \left(\frac{hc}{\lambda'}, \frac{hc}{\lambda'}\cos\theta, \frac{hc}{\lambda'}\sin\theta, 0\right), \quad p'_e = \left(mc^2, m\mathbf{v}\right). \tag{4.84}$$

Interestingly, we shall not use p'_e. Now, using the conservation of energy and momentum:

$$p_\gamma + p_e = p'_\gamma + p'_e$$

$$\Rightarrow p'^2_e = \left(p_\gamma + p_e - p'_\gamma\right)^2$$

$$\Rightarrow p'^2_e = p_\gamma{}^2 + p_e{}^2 + p'^2_\gamma + 2p_e \cdot (p_\gamma - p'_\gamma) - 2p_\gamma \cdot p'_\gamma$$

$$\Rightarrow m_e{}^2 = 0 + m_e{}^2 + 0 + 2m_e c^2 \left(\frac{hc}{\lambda} - \frac{hc}{\lambda'}\right)$$

$$- 2\frac{hc}{\lambda}\frac{hc}{\lambda'}(1 - \cos\theta) \tag{4.85}$$

$$\Rightarrow 2m_e c^2 \left(\frac{hc}{\lambda} - \frac{hc}{\lambda'}\right) = 2\frac{hc}{\lambda}\frac{hc}{\lambda'}(1 - \cos\theta)$$

$$\Rightarrow m_e c^2 \left(\frac{1}{\lambda} - \frac{1}{\lambda'}\right) = \frac{hc}{\lambda\lambda'}(1 - \cos\theta)$$

$$\Rightarrow \lambda' - \lambda = \frac{h}{m_e c}(1 - \cos\theta)$$

This shows the elegance of the 4-vector method over the traditional method in solving kinematic problems.

4.3 Invariants: Mandelstam Variables

Consider a two-body final state scattering process: $A + B \rightarrow C + D$. Here, one expects two independent kinematic variables; namely, (i) the incident energy and (ii) the scattering angle. Can one then express the invariant amplitude for the above process in terms of the Lorenz invariant variables? The answer is yes, and we have got the particle four-momenta in hand to construct the Lorentz scalars- $p_A \cdot p_B$, $p_A \cdot p_C$, $p_A \cdot p_D$. We have $p_i^2 = m_i^2$ and the energy-momentum conservation gives $p_A + p_B = p_C + p_D$. This leaves out only two of the three variables as independent. Although from an experimental point of view, the energy and the scattering angle are more convenient to use, theoretically, it is conventional to use the related Mandelstam variables. These are:

$$s \equiv (p_A + p_B)^2,$$
$$t \equiv (p_A - p_C)^2, \tag{4.86}$$
$$u \equiv (p_A - p_D)^2,$$

Some of the important properties of the Mandelstam variables are:

1. It is easy to show that $s + u + t = \sum m_i^2$, where m_i is the rest mass of the ith particle.
2. s is the square of the total center-of-mass energy of the process $AB \rightarrow CD$.
3. The crossed channel reactions $A\bar{D} \rightarrow C\bar{B}$ and $\bar{D}B \rightarrow C\bar{A}$ are called the u and t channels, respectively. This is because u and t are equal to the total center-of-mass energy in the respective channels.
4. For a "s-channel process", s is the square of the center-of-mass energy, t is the squared 4-momentum transfer, and u has no physical meaning (is the crossed momentum transfer squared), as there is no Lorentz system, where it reduces to something obvious.

Example 4.5   Find the center-of-mass energy of A, in terms of the Mandelstam variables and the particle masses.

Solution 4.5

$$\begin{aligned}
s &= (p_A + p_B)^2 \\
&= p_A^2 + p_B^2 + 2p_A \cdot p_B \\
&= p_A^2 + p_B^2 + 2(E_A E_B - \mathbf{p}_A \cdot \mathbf{p}_B)
\end{aligned}$$

In the CM frame, $\mathbf{p}_A \cdot \mathbf{p}_B = -\mathbf{p}_A^2$.

$$\therefore \; p_A^2 = E_A^2 - \mathbf{p}_A^2 \tag{4.87}$$
$$\Longrightarrow \mathbf{p}_A^2 = E_A^2 - p_A^2 = -\mathbf{p}_A \cdot \mathbf{p}_B \tag{4.88}$$

and $p_B^2 = E_B^2 - \mathbf{p}_B^2 = E_B^2 - \mathbf{p}_A^2$ ($\because |\mathbf{p}_A| = |\mathbf{p}_B|$ in CM system)

$$\Longrightarrow E_B = \sqrt{p_B^2 + \mathbf{p}_A^2}$$
$$= \sqrt{p_B^2 + E_A^2 - p_A^2} \quad \text{[by using Eq. 4.87]} \tag{4.89}$$

Now $s = p_A^2 + p_B^2 + 2[E_A\sqrt{p_B^2 + E_A^2 - p_A^2} + E_A^2 - p_A^2]$ (by using Eqs. 4.88 and 4.89).

$$s + p_A^2 - p_B^2 - 2E_A^2 = 2E_A\sqrt{p_B^2 + E_A^2 - p_A^2}$$

Squaring both sides of the above equation, we get:

$$(s + p_A^2 - p_B^2)^2 + 4E_A^4 - 4E_A^2(s + p_A^2 - p_B^2) = 4E_A^2(p_B^2 + E_A^2 - p_A^2)$$

$$\implies (s + p_A^2 - p_B^2)^2 = 4E_A^2(p_B^2 - p_A^2 + s + p_A^2 - p_B^2)$$

$$= 4E_A^2 s$$

$$\implies E_A^2 = \frac{(s + p_A^2 - p_B^2)^2}{4s}$$

$$\implies E_A^{CM} = \frac{(s + m_A^2 - m_B^2)}{2\sqrt{s}}$$

Example 4.6 Find the lab (assuming B is at rest) energy of A, in terms of the Mandelstam variables and the particles masses.

Solution 4.6 In the laboratory system, $\mathbf{p_B} = 0$ and $E_B = m_B$.

$$s = (p_A + p_B)^2$$

$$= p_A^2 + p_B^2 + 2(E_A E_B - \mathbf{p_A} \cdot \mathbf{p_B}) \quad (\because \mathbf{p_B} = 0)$$

$$= m_A^2 + m_B^2 + 2(E_A m_B)$$

$$\implies E_A^{lab} = \frac{s - m_A^2 - m_B^2}{2m_B}$$

Example 4.7 Find the total center-of-mass energy ($E_{TOT} = E_A + E_B = E_C + E_D$).

Solution 4.7 In the center-of-mass frame,

$$(p_A + p_B)^2 = (E_A + E_B)^2 - (\mathbf{p_A} + \mathbf{p_B})^2$$

$$= E_{TOT}^2$$

$$\implies s = E_{TOT}^2$$

$$\implies E_{TOT}^{CM} = \sqrt{s}.$$

Example 4.8 For elastic scattering of identical particles like the Moller scattering ($e^- e^- \to e^- e^-$), show that the Mandelstam variables become

$$s = 4(\mathbf{p}^2 + m^2)$$

$$t - 2\mathbf{p}^2(1 - \cos\theta)$$

$$u - 2\mathbf{p}^2(1 + \cos\theta)$$

Here $\mathbf{p}$ is the CM momentum of the incident particle and θ is the scattering angle.

Solution 4.8 We know $s = (p_A + p_B)^2 = (E_A + E_B)^2 - (\mathbf{p_A} + \mathbf{p_B})^2$.
In CM frame, $\mathbf{p_A} + \mathbf{p_B} = 0$ and $E_A = E_B = \sqrt{\mathbf{p}^2 + m^2}$

$$s = (2E_A)^2$$
$$= 4(p^2 + m^2)$$

Further, by definition,

$$t = (p_A - p_C)^2$$
$$= (E_A - E_C)^2 - (\mathbf{p_A} - \mathbf{p_C})^2$$

But $E_A = E_C$ and

$$(\mathbf{p_A} - \mathbf{p_C})^2 = \mathbf{p_A}^2 + \mathbf{p_C}^2 - 2\mathbf{p_A} \cdot \mathbf{p_C}$$
$$= 2\mathbf{p}^2(1 - \cos\theta)$$
$$\Longrightarrow t = -2\mathbf{p}^2(1 - \cos\theta)$$

Again, as per the definition of u,

$$u = (p_A - p_D)^2$$
$$= (E_A - E_D)^2 - (\mathbf{p_A} - \mathbf{p_D})^2$$
$$= -(\mathbf{p_A}^2 + \mathbf{p_D}^2 - 2\mathbf{p_A} \cdot \mathbf{p_D}) \quad (\because E_A = E_D)$$
$$= -2\mathbf{p}^2(1 + \cos\theta)$$
$$\Longrightarrow u = -2\mathbf{p}^2(1 + \cos\theta)$$

4.4   Estimation of Threshold Energy

The threshold energy for any interaction is the energy required to produce the final state particles (any number) with zero kinetic energy. Or in other words, it is the projectile energy in the laboratory system, so as to produce the final state particles. In some of the books, the threshold energy is also defined as the minimum kinetic energy of the incident beam so that the final-state particles are just produced with zero kinetic energy.

Consider an interaction, where the projectile A with mass, m_A and energy in the laboratory system, E_{lab}, collides with a target, B of mass, m_B, at rest

($\mathbf{p}_B = 0$). In this interaction, there are n final-state particles produced. Let them be: $C_1, C_2, ...C_n$, with masses, $m_{C_1}, m_{C_2}, ..., m_{C_n}$, respectively.

$$A + B \rightarrow C_1 + C_2 + ... + C_n$$

To estimate the threshold energy for this general interaction, we first write the four-momenta for individual particles to calculate the total four-momenta before and after the collisions, suitably choosing the frame of reference, which should be clear in a moment. Then we apply the Lorentz invariant property of the scalar product of the four-momenta to solve the problem. Let's now write the four-momenta.

For particle A: $p_A = (E_A, \mathbf{p}_A)$, for particle B: $p_B = (m_B, \mathbf{0})$, and for the final state particles, as their kinetic energy is zero at the threshold production, the four-momenta is: $(m_{C_1} + m_{C_2} + ... + m_{C_n}, \mathbf{0})$.

In the laboratory frame, before the collision,

$$p^{\mu}_{TOT} = (E_A + m_B, \mathbf{p}_A)$$

In the center-of-mass frame, after the collision,

$$p^{\mu}_{TOT'} = \left(\sum_{i=0}^{n} m_{C_i}, \mathbf{0} \right) \equiv (M_C, \mathbf{0})$$

As p^2_{TOT} after and before the collision is Lorentz invariant,

$$p^{\mu}_{TOT}\, p_{\mu,TOT} = p^{\mu}_{TOT'}\, p_{\mu,TOT'}$$

$$\Rightarrow (E_A + m_B)^2 - \mathbf{p}_A^2 = M_C^2$$

$$\Rightarrow E_A^2 + m_B^2 + 2E_A m_B - \mathbf{p}_A^2 = M_C^2$$

$$\Rightarrow m_A^2 + m_B^2 + 2E_A m_B = M_C^2$$

$$\Rightarrow E_A = \frac{M_C^2 - m_A^2 - m_B^2}{2m_B}$$

$$\Rightarrow \boxed{E^{Th}_{lab} = \frac{\left(\sum_{i=1}^{n} m_{C_i}\right)^2 - m_A^2 - m_B^2}{2m_B}} \tag{4.90}$$

This is the formula for calculating the threshold energy of an interaction for producing any number of final-state particles. Let's now apply this to a real problem.

Example 4.9 Threshold energy for $\eta\prime$ production in WASA-COSY experiment at Jülich, Germany.

In the WASA-COSY experiment at Jülich, Germany, a proton beam is used on a fixed proton target to produce $\eta\prime$ mesons. Estimate the threshold energy required for the formation of $\eta\prime$: $(pp \rightarrow pp\eta\prime)$.

Solution 4.9 Given masses of the particles are, proton mass, $m_p = 938.27$ MeV and $\eta\prime$ mass, $m_{\eta\prime} = 957.78$ MeV. Using Eq. 4.90, the threshold energy for $\eta\prime$ production is given by

$$
\begin{aligned}
E_{lab}^{Th} &= \frac{(\sum_{i=1}^{n} m_{c_i})^2 - m_A^2 - m_B^2}{2m_B} \\[2mm]
&= \frac{(2m_p + m_{\eta\prime})^2 - 2m_p^2}{2m_p} \\[2mm]
&= \frac{[2 \times 938.27 + 957.78]^2 - 2 \times (938.27)^2}{2 \times 938.27} \\[2mm]
&= 3342.67 \text{ MeV} \simeq 3.34 \text{ GeV}
\end{aligned}
$$

It would be interesting to calculate the velocity of the projectile proton to have a feeling of the relativistic energy we are working with. We know $E = m\gamma$, where $\gamma = \frac{1}{\sqrt{1-\beta^2}}$, is the Lorentz factor.

$\gamma = E/m = 3342.67/938.27 = 3.562$. Solving for the velocity, we get $\beta = 0.95978$, using $\beta = (1 - \frac{1}{\gamma^2})^{\frac{1}{2}}$. This means the projectile proton travels with 95.97835% of the speed of light. What is your guess for the speed of the proton at the Large Hadron Collider, CERN, where the beam energy is 3.5 TeV? If you estimate this, you shall find that the beam proton has a Lorentz factor ~ 3730.27 and it travels with 99.9999964% the speed of light. Now you may be surprised to see that even if the beam energy is almost 1000 times higher, the speed of the proton doesn't change proportionately. What could be the reason? We shall discuss it further in the subsequent chapters. And you will realize that, with a change of energy and consequently the speed of the accelerated particle, the relativistic mass of the particle changes, and it becomes difficult to accelerate further.

You may wonder to know that the *threshold energy for an anti-particle production is higher* than that required for the production of a particle! The first anti-proton was produced in the laboratory at Berkley Bevatron in the reaction: $p + p \rightarrow p + p + p + \bar{p}$. A high-energy proton strikes a proton at rest to produce a pair of proton and anti-proton, along with the original particles. At first sight, one can infer that conserving baryon number, we have two new particles in the final state. This will also demand a higher threshold energy. Let's now estimate the threshold energy for this reaction to have a clear view of the discussed picture.

Example 4.10 Anti-proton creation in Bevatron.

Solution 4.10 The idea to create antiprotons in the laboratory was to strike a high-energy proton (beam energy E_p) with a proton at rest, creating a proton-antiproton pair in addition to the original particles. The interaction, following the conservation laws (baryon number and charge) is the following:

$$p + p \rightarrow p + p + p + \bar{p}$$

For the estimation of the threshold energy for this interaction, we need to find the minimum energy, E_p, if the final state particle are just produced without any kinetic energy. To solve this problem, one can proceed as follows:

(i) Write the 4-momenta before and after the interaction,
(ii) Take the 4-vector scalar product before and after the collision, which is Lorentz invariant,
(iii) Solve the equation, making these two equal, and you will get the required E_p.

The total energy-momentum 4-vector, P^{μ}_{TOT} is being a conserved quantity, is independent of where it is evaluated, i.e., either in the laboratory frame or in the CM-frame. Let's do it in the laboratory frame for ease of calculation. Now the total 4-momenta before the collision in the laboratory frame is:

$$P^{\mu}_{\text{TOT}} = \left(E_p + m, |\mathbf{p}|\right)$$

Here m is the mass of a proton, which is equal to the mass of an antiproton because of the CPT invariance. $\mathbf{p}$ is the three momentum of the incident proton. We now proceed to write the energy-momentum 4-vector for the final state particles in the CM-frame, $P^{\mu}_{\text{TOT}}{}'$. As the particles are produced at the threshold, they are at rest (zero kinetic energy).

$$P^{\mu}_{\text{TOT}}{}' = (4m, |\mathbf{0}|)$$

Taking the invariant products:

$$P_{\mu\text{TOT}} P^{\mu}_{\text{TOT}} = P_{\mu\text{TOT}}{}' P^{\mu}_{\text{TOT}}{}'$$

$$\Rightarrow (E_p + m)^2 - \mathbf{p}^2 = (4m)^2$$

$$\Rightarrow (E_p^2 - |\mathbf{p}|^2) + 2E_p m + m^2 = 16m^2$$

$$\Rightarrow 2E_p m + 2m^2 = 16m^2$$

$$\Rightarrow E_p - 7m$$

The threshold energy for the production of an antiproton is seven times its rest mass. In other words, as the kinetic energy equals the energy of the particle minus the rest mass energy, in this example, the minimum kinetic energy required for the incident proton to produce an antiproton is 6 m.

Example 4.11 B-meson creation in KEKB: KEKB accelerator copiously produces B-mesons, especially for the study of CP-violation. If B-mesons are produced from $e^+ e^- \rightarrow \Upsilon(4S)$, find the center-of-mass energy required for B-meson production.

Solution 4.11 B-mesons are produced from the interaction: $e^+ e^- \rightarrow \Upsilon(4S)$. $\Upsilon(4S)$ resonance is a bound state of $b\bar{b}$, with rest mass, $m_{\Upsilon(4S)} = 10.58$ GeV. $\Upsilon(4S)$ decays to $B^0 \bar{B}^0$, $B^+ B^-$ with a branching fraction of around 96%. As the rest mass of $\Upsilon(4S)$ is 10.58 GeV, this energy corresponds to the threshold energy, and this is the center-of-mass energy required to produce B-mesons. KEKB is an asymmetric electron-positron collider with a center-of-mass energy of 10.58 GeV.

Kinematics in Heavy-Ion Collisions 5

The study of quark-gluon plasma in heavy-ion collisions is a unique window into the early universe and the fundamental nature of matter.

— *David J. Gross*

After the discovery of the quark structure of nucleons through the deep inelastic (DIS) scattering (*ep*) in the Stanford Linear Accelerator (SLAC), the search for a deconfined state of quarks and gluons started with the Alternating Gradient Synchrotron (AGS) at the Brookhaven National Laboratory (BNL), New York, USA; Super Proton Synchrotron (SPS) at the European Laboratory for Nuclear Research (CERN), Switzerland; Relativistic Heavy Ion Collider (RHIC) at BNL, and the Large Hadron Collider (LHC) at CERN. AGS and SPS were fixed-target experiments, whereas the collider era in heavy-ion collisions with ultra relativistic energies started with RHIC getting operational in the year 2000. Though heavy-ion collisions with a moderate energy of around 1–2 GeV/nucleon were done at Bevalac in Lawrence Berkeley National Laboratory in the 1970s. Some of the future facilities, like the Facility for Antiproton and Ion Research (FAIR) at Darmstadt, Germany, Nuclotron-based Ion Collider fAcility (NICA) at JINR Dubna, Russia, are planned to be fixed-target machines. In this chapter, we plan to discuss the kinematics required for understanding heavy-ion collisions in particular, and high-energy collisions in general. Before this, let us try to motivate ourselves for the necessity of heavy-ion collisions from the perspective of the physics of the Standard Model. Figure 5.1 shows the schematic view of the CERN Large Hadron Collider accelerator complex, which is at present the largest particle accelerator on Earth. The LHC in Run 3 is colliding protons at $\sqrt{s} = 13.6\,\text{TeV}$, heavy-ions (Pb-Pb) and light-ions (O-O and Ne-Ne) at $\sqrt{s_{NN}} = 5.36\,\text{TeV}$, along with p-O at $\sqrt{s_{NN}} = 9.62\,\text{TeV}$.

© The Author(s), under exclusive license to Springer Nature Switzerland AG 2025   101
R. Sahoo, *Relativistic Kinematics*, Lecture Notes in Physics 1046,
https://doi.org/10.1007/978-3-032-09510-7_5

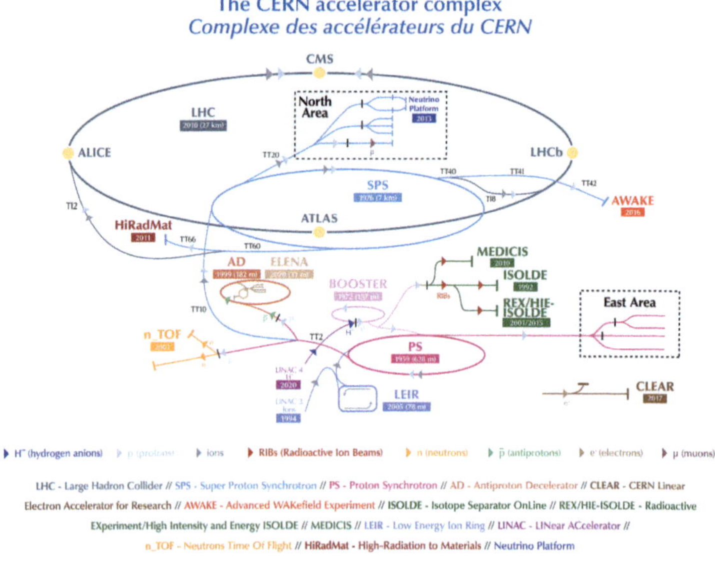

Fig. 5.1 Schematic of the CERN accelerator complex. Reprinted with permission from [4]. ©2022, CERN. All rights reserved

Why Heavy-ion Collisions?

The basic aim of colliding heavy ions is to create a plasma of the fundamental constituents of matter, quarks and gluons (collectively known as *partons*). This name was given by Richard Feynman to the fundamental constituents of matter. He combined the words *"part"* and *"-on"* to describe the point-like constituents of protons and neutrons, which were later discovered to be quarks and gluons. This short-lived plasma, which is believed to have been produced in the very infancy of our Universe, had a very high temperature, energy density, and expanded through extremely high pressure, sometimes reaching the collective constituent velocities around 70% of the speed of light at the freeze-out stage. Heavy ion collision gives an opportunity to create such a plasma in the laboratory using modern accelerators like the LHC at CERN. This facilitates the study of the early Universe phenomena and their spacetime evolution, thereby giving a solid ground of critical understanding of the creation and evolution of the Universe. Heavy ion collisions also serve as a test bench for verifying the confinement-deconfinement phase transition of QCD, otherwise called as quark-hadron phase transition. By changing the collision energy and colliding species in heavy-ion collisions, physicists try to scan the QCD phase diagram, which is rich in physics.

Figure 5.2 shows how the history of discovery on the energy frontier has been enabled by the history of invention (red arrows) in accelerator science and technology. This is the famous Livingston plot. With the discovery of radioactivity, it was Rutherford who first used the constituents of matter, like the α, β and γ

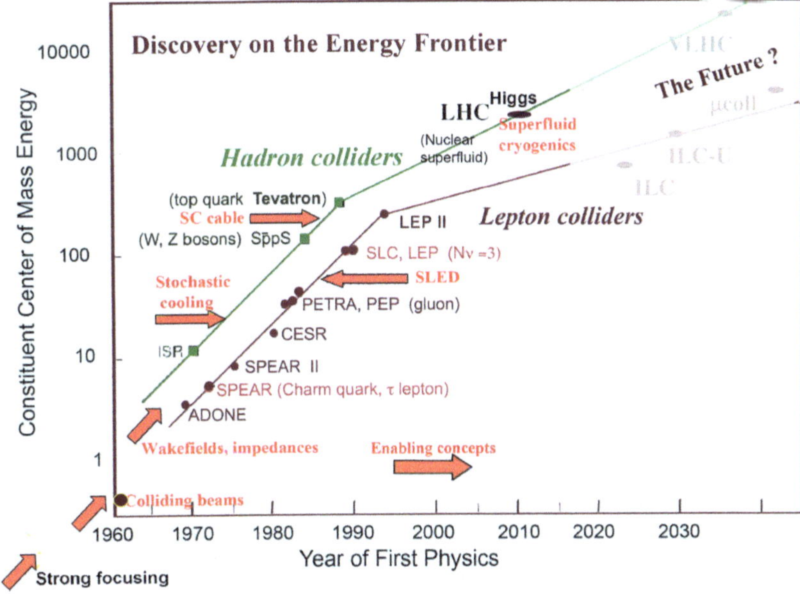

Fig. 5.2 The Livingston plot: Collision energy vs. the year of first physics. Reprinted with permission from [5]. ©2014, The Author(s)

particles/rays, to probe the building blocks of matter. That was better demonstrated through his discovery of the atomic nucleus by the α-scattering experiment. However, he recognized that radioactive sources have extremely limited capability to be used to probe the inner structure of matter. Since then, with the advent of accelerators and the enabling technologies, accelerator-based experiments have been the mainstay and driving force of experimental high-energy physics. Through the energy and luminosity frontiers, the modern-day accelerators have enabled the particle physics discoveries in an unprecedented way. Along with the discovery of the most elusive Pauli's neutrinos in 1956, the 2012 discovery of the massive Higgs boson, the so-called *"God Particle"* by the CMS and ATLAS experiments at CERN, is a testimony to this.

5.1   Collider vs Fixed Target Experiment

As depicted in Fig. 5.3, a fixed target experiment refers to a laboratory system/frame (LS) and a collider experiment refers to the center-of-mass(momentum) system (CMS). While each system has its own advantages and disadvantages in terms of specific physics goals, we shall now discuss some of the kinematic aspects of LS and CMS.

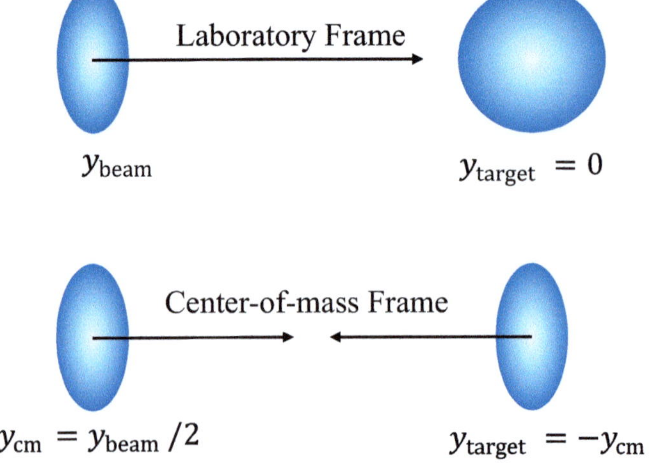

Fig. 5.3 A schematic showing a two-body collision in the laboratory frame/system (LS) and collision in the center-of-mass (CMS) system

5.1.1   For Symmetric Collisions ($A + A$)

Consider the collision of two particles. In LS, the projectile with momentum $\mathbf{p}_1$, energy E_1 and mass m_1 collides with a particle of mass m_2 at rest. The 4-momenta of the particles are

$$p_1 = (E_1, \mathbf{p}_1), \qquad p_2 = (m_2, \mathbf{0})$$

In CMS, the momenta of both the particles are equal and opposite, the 4-momenta are

$$p_1^* = (E_1^*, \mathbf{p}_1^*), \qquad p_2^* = (E_2^*, -\mathbf{p}_1^*)$$

We use asterisk (*) for variables in the CM frame to distinguish them from the laboratory frame. The total 4-momentum of the system is a conserved quantity in the collision.

In CMS,

$$p_\mu p^\mu = (p_1{}^* + p_2{}^*)^2 = (E_1{}^* + E_2{}^*)^2 - (\mathbf{p}_1{}^* + \mathbf{p}_2{}^*)^2$$
$$= (E_1{}^* + E_2{}^*)^2$$
$$= E_{cm}^2 \equiv s \qquad\qquad (5.1)$$

$\sqrt{s}$ is the total energy in the CMS, which is the invariant mass of the CMS.

In LS,

$$p_\mu p^\mu = (p_1 + p_2)^2 = m_1^2 + m_2^2 + 2E_1 m_2 \tag{5.2}$$

Hence

$$\boxed{E_{cm} = \sqrt{s} = \sqrt{m_1^2 + m_2^2 + 2E_{proj} m_2}} \tag{5.3}$$

where $E_1 = E_{proj}$, the projectile energy in LS. Hence, it is evident here that the CM frame with an invariant mass $\sqrt{s}$ moves in the laboratory in the direction of $\mathbf{p}_1$ with a velocity corresponding to:
Lorentz factor,

$$\gamma_{cm} = \frac{E_1 + m_2}{\sqrt{s}} \tag{5.4}$$

$$\Rightarrow \sqrt{s} = \frac{E_{lab}}{\gamma_{cm}}, \tag{5.5}$$

this is because $E = \gamma m$. The center-of-mass or center of momentum frame (CM/CMS) is at rest, and the total momentum is zero. This makes it a suitable choice for solving kinematics problems.

Note We know that for a collider with a head-on collision ($\theta = 180^0$)

$$s = E_{cm}^2 = m_1^2 + m_2^2 + 2(E_1.E_2 - \mathbf{p}_1.\mathbf{p}_2)$$
$$= m_1^2 + m_2^2 + 2(E_1.E_2 + |\mathbf{p}_1||\mathbf{p}_2|)$$

For relativistic collisions, $m_1, m_2 \ll E_1, E_2$

$$E_{cm}^2 \simeq 4E_1 E_2 \tag{5.6}$$

For two beams crossing at an angle θ,

$$\boxed{E_{cm}^2 = 2E_1 E_2(1 - \cos\theta)} \tag{5.7}$$

In the limit, $m_1, m_2 \ll E_1, E_2$, energies become equal to the corresponding momenta. The CM energy available in a collider with equal beam energy (E) for a new particle production rises linearly with E i.e.

$$E_{cm} \sim 2E \tag{5.8}$$

For a fixed-target experiment, the CM energy rises as the square root of the incident energy:

$$E_{cm} \simeq \sqrt{2m_2 E_1} \tag{5.9}$$

Hence, the highest energy available for new particle production is achieved at collider experiments. For example, at the SPS fixed-target experiment to achieve a CM energy of 17.3 AGeV, the required incident beam energy is 158 AGeV.

Problem Suppose two identical particles, each with mass m and kinetic energy T, collide head-on. What is their relative kinetic energy, T' (i.e. K.E. of one in the rest frame of the other). Apply this to an electron-positron collider, where the kinetic energy of the electron (positron) is 1 GeV. Find the kinetic energy of the electron if the positron is at rest (fixed target). Which experiment is preferred, a collider or a fixed-target experiment?

Note Most of the time, the energy of the collision is expressed in terms of nucleon-nucleon center-of-mass energy. In the nucleon-nucleon CM frame, two nuclei approach each other with the same boost factor γ. The nucleon-nucleon CM is denoted by $\sqrt{s_{NN}}$ and is related to the total CM energy by

$$\sqrt{s} = A \sqrt{s_{NN}} \tag{5.10}$$

This is for a symmetric collision with a number of nucleons in each nucleus as A. The colliding nucleons approach each other with energy $\sqrt{s_{NN}}/2$ and with equal and opposite momenta. The rapidity of the nucleon-nucleon center of mass is $y_{NN} = 0$ and taking $m_1 = m_2 = m_p$, the projectile and target nucleons are at equal and opposite rapidities (Fig. 5.3).

$$y_{proj} = -y_{target} = \cosh^{-1} \frac{\sqrt{s_{NN}}}{2m_p} = y_{beam}. \tag{5.11}$$

Note Lorentz Factor

$$\gamma = \frac{E}{M} = \frac{\sqrt{s}}{2A\,m_p}$$

$$= \frac{A\sqrt{s_{NN}}}{2A\,m_p} = \frac{\sqrt{s_{NN}}}{2\,m_p}$$

$$= \frac{E_{beam}^{CMS}}{m_p} \tag{5.12}$$

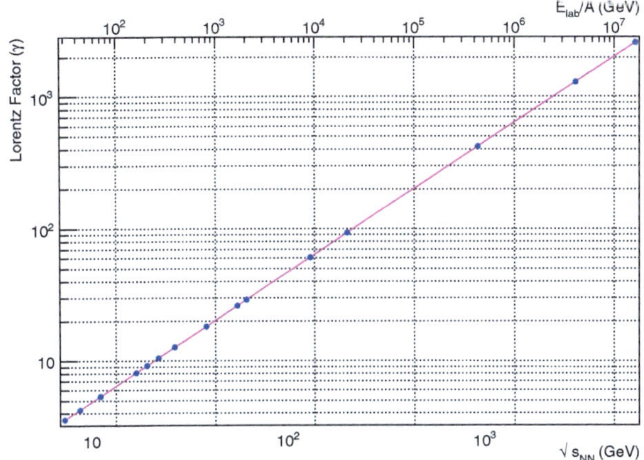

Fig. 5.4 A plot to demonstrate how the Lorentz Factor increases with collision energy for a symmetric collision. The center-of-mass system is compared with fixed target experiment

where E and M are energy and mass in CMS, respectively. Assuming the mass of a proton, $m_p \sim 1\,\text{GeV}$, the Lorentz factor is of the order of the beam energy in CMS for a symmetric collision. The variation of Lorentz factor as a function of center-of-mass energy is shown in Fig. 5.4.

Example 5.1 (KEKB Asymmetric Collider) The KEKB in Japan is an asymmetric e^+e^- collider with beam energy of e^- 8.0 GeV and e^+ beam is having energy 3.5 GeV. The two beams collide at an angle of ± 11 milli-radian (mRad). Estimate the center-of-mass energy.

Solution 5.1 Given the beam energies of e^-, $E_1 = 8\,\text{GeV}$ and for e^+, $E_2 = 3.5\,\text{GeV}$. The angle of collision, $\theta = \pm 11$ mRad.
   Now, $1\,\text{mRad} \times \frac{180}{1000\pi} = 0.0573°$. Hence, $\theta = 11 \times 0.0573 = 0.6303°$. Using Eq. (5.7),

$$E_{cm}^2 = 2E_1E_2(1 - \cos\,\theta)$$

$$= 2E_1E_2[1 - \cos\,(\pi - \theta)]$$

$$= 2E_1E_2(1 + \cos 0.6303)$$

$$= 4E_1E_2$$

$$\Rightarrow E_{cm} = 2\sqrt{E_1E_2}$$

$$\Rightarrow E_{cm} = 2 \times \sqrt{8 \times 3.5}\,\text{GeV}$$

$$= 10.58\,\text{GeV}$$

It is interesting to note that such an asymmetric energy collision with collision angle of ± 11 mRad is chosen to optimize the cost and material budget for the detector system, as the final state particles will be produced in a certain forward solid angle only, instead of being produced in a 4π geometry, as is expected for a collider with 180^0 collision angle.

5.1.2   For Asymmetric Collisions $(A + B)$

Usually, in an accelerator, depending on the accelerator parameters, to accelerate a single charged state (for example, a proton), the highest energy achievable or that to for a particular physics purpose of a baseline study for heavy-ion collisions, one aims to run the accelerator to obtain a given center-of-mass energy for $p + p$ collisions. In that case, taking this energy for $p + p$ collisions, if we need to have symmetric or asymmetric heavy-ion collisions, one should keep the following physics considerations in mind. In a heavy-ion collision, the atom is fully stripped off with electrons, and the nucleus has Z number of protons and N number of neutrons $(A - Z)$. Neutral particles can't be accelerated independently. The charged state (fully stripped means the charged state equals the total number of protons inside the nucleus) of the nucleus is responsible for obtaining the required energy in the acceleration process. On the other hand, the neutrons inside the nucleus are just *"free riders"* in the process, and we need to spend the energy of the protons to accelerate the neutrons. This means the center-of-mass energy for heavy-ion collisions is just not $Z - times$ of the $p + p$ collision energy, but rather a reduced value. Let's now estimate the former energy, given the $p + p$ center-of-mass energy.

Using Eq. (5.6), for head-on heavy-ion collisions,

$$\sqrt{s_{NN}} \simeq 2\sqrt{E_1 E_2} \tag{5.13}$$

For a collision of two different nuclei with charges (atomic number) Z_1, Z_2 and atomic masses A_1, A_2, respectively, $E_1 = E_{proton}(Z_1/A_1)$ and $E_2 = E_{proton}(Z_2/A_2)$. The above formula with these substitutions becomes:

$$\sqrt{s_{NN}} \simeq 2E_{proton}\sqrt{\frac{Z_1 Z_2}{A_1 A_2}} \tag{5.14}$$

$$\Rightarrow \boxed{\sqrt{s_{NN}} \simeq \sqrt{s_{pp}}\sqrt{\frac{Z_1 Z_2}{A_1 A_2}}}. \tag{5.15}$$

Here $\sqrt{s_{pp}}$ is the center-of-mass energy of $p+p$ collision. For symmetric collisions with atomic number and mass– (Z, A), this formula becomes

$$\sqrt{s_{\mathrm{NN}}} \simeq \sqrt{s_{\mathrm{pp}}}\,\frac{Z}{A}. \tag{5.16}$$

Example 5.2 The Large Hadron Collider collides $p + p$, $p+$Pb ($_{82}\mathrm{Pb}^{208}$), Xe+Xe ($_{54}\mathrm{Xe}^{131}$) and Pb+Pb at different center-of-mass energies. Given the proton beam energies (or $p + p$ center-of-mass energies) in a particular accelerator setup, find the corresponding center-of-mass energies achievable for $p+$Pb, Xe+Xe and Pb+Pb collisions.

Solution 5.2 Using Eq. (5.15), the center-of-mass energy for $p+$A collision is given by (taking $Z_1 = A_1 = 1$ for proton beam)

$$\sqrt{s_{\mathrm{pA}}} \simeq \sqrt{s_{\mathrm{pp}}}\sqrt{\frac{Z_2}{A_2}}. \tag{5.17}$$

Using Eqs. (5.15)–(5.17), one can estimate the center-of-mass energies in $p+$A and A+A collisions, given the corresponding beam/center-of-mass energies in $p + p$ collisions. The estimations are tabulated in Table 5.1 based on LHC $p + p$ data taking. The (*)-marked energies for $p+$A and A+A collisions correspond to real LHC data collected till August 2025.

Note that LHC has taken data for O+O and Ne+Ne at $\sqrt{s_{\mathrm{NN}}} = 5.36\,\mathrm{TeV}$ for which the corresponding center-of-mass-energy for $p + p$ collisions is $\sqrt{s_{pp}} = 10.72\,\mathrm{TeV}$. This is excluded from the table, as the LHC did not collide protons at $\sqrt{s_{pp}} = 10.72\,\mathrm{TeV}$.

Let's now discuss the rapidity shift in asymmetric collision systems. This is given by

$$\Delta y \simeq \frac{1}{2}\,\ln\left[\frac{Z_1 A_2}{Z_2 A_1}\right] \tag{5.18}$$

Table 5.1 Table of collision energy, $\sqrt{s_{\mathrm{NN}}}$ for $p+$A and A+A collisions at the LHC, as estimated from the corresponding $p + p$ data taken till August 2025. The center-of-mass energies with *-marks represent data already taken

$\sqrt{s_{pp}}$ (TeV)	0.9*	2.36*	2.76*	5.02*	7 *	8*	13*	13.6*
$p+$Pb	0.57	1.48	1.74	3.16	4.4	5.02*	8.16*	
Pb+Pb	0.35	0.93	1.09	1.98	2.76*	3.17	5.14	5.36*
Xe+Xe	0.37	0.97	1.14	2.07	2.88	3.3	5.44*	
O+O/Ne+Ne	.	.	.	.	.	.	.	6.8
p+O	.	.	.	.	.	.	.	9.62*

This is because the center-of-mass frame of the pA collision doesn't coincide with the laboratory center-of-mass frame. The rapidity shift in $p + Pb$ collision is

$$\Delta y \simeq \frac{1}{2} \ln \left[\frac{Z_1 A_2}{Z_2 A_1} \right]$$

$$= \frac{1}{2} \ln \left[\frac{82 \times 1}{1 \times 208} \right]$$

$$= -0.465$$

Hence $\Delta y = \pm 0.465$ for $p + Pb$ collisions, flipping the beams. This rapidity shift needs to be taken into account for the comparison with Pb+Pb data.

At LHC, the maximum proton beam energy is 7 TeV, while the maximum Pb beam energy is 2.75 TeV. The difference in available energy is due to the charge-to-mass ratio, Z/A. The more is the number of neutrons in the nucleus, difficult it is to accelerate to higher energies. Because of different energies, the two beams will also not have the same rapidity. For the proton beam $y_p = 9.61$ and for the Pb beam it is $y_{Pb} = 8.67$. Thus the center of the collision is shifted away from $y_{cm} = 0$ by $\Delta y_{cm} = (y_p - y_{Pb})/2 = 0.47$.

5.1.3   The Energy and Velocity of the Center of Momentum

As discussed in Ref. [6], let us consider a Lorentz system- call it the laboratory system (frame), with two particles with masses, m_1 and m_2 and four momenta p_1 and p_2, respectively as shown in Fig. 5.5.

What is the centre of mass energy, E ?

1. It is independent of the Lorentz system, where p_1 and p_2 are defined.
2. It must be possible to have an answer in terms of the three invariants, namely $p_1^2 = m_1^2$ and $p_2^2 = m_2^2$ and $[p_1 p_2$ or $(p_1 + p_2)^2$ or $(p_1 - p_2)^2]$.

The answer would be trivial in the center of momentum (mass) frame itself.

In CM frame,

$$(\mathbf{p}_1^* + \mathbf{p}_2^*) = 0$$

$$\Longrightarrow p_1 + p_2 = (E_1^* + E_2^*, \mathbf{0})$$

$$\text{and } E^* = E_1^* + E_2^* \tag{5.19}$$

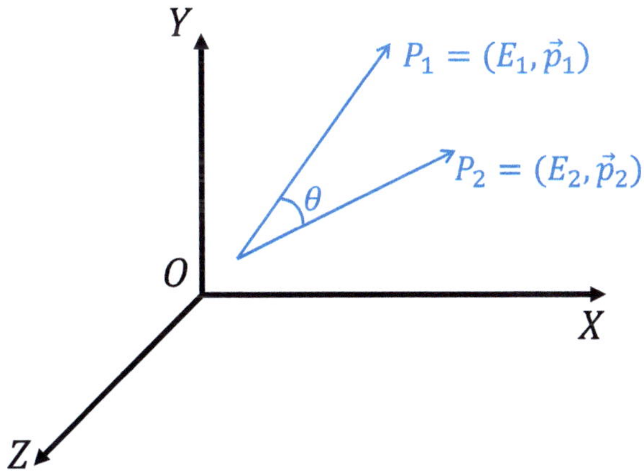

Fig. 5.5 The Laboratory system: two particle kinematics showing the 4-momenta of two particles

Hence,

$$\begin{aligned}
(E^*)^2 &= (E_1^* + E_2^*)^2 \\
&= (p_1^* + p_2^*)^2 \\
&= (p_1 + p_2)^2.
\end{aligned} \tag{5.20}$$

This is because $(p_1 + p_2)^2$ is Lorentz invariant.

Let us define the total mass M of the system as the square of the total 4-momentum as:

$$\boxed{M^2 = (p_1 + p_2)^2 = P^2 = (E^*)^2 = (E_1 + E_2)^2 - (\mathbf{p}_1 + \mathbf{p}_2)^2 = Invariant}$$

Kinematically, for the given system, the two particles p_1 and p_2 are equivalent to one single particle with 4-momentum P and mass $M = E_{CM}$. Generalizing this, one can consider these individual particles to represent a system of particles.

Further, we know

$$\boxed{\mathbf{p} = m\mathbf{v}\gamma, \text{ and } E = m\gamma}, \text{ where } \gamma = \frac{1}{\sqrt{1-\beta^2}}.$$

Hence, the 4-momentum of the two-particle system is given by:
$\mathbf{P} = M\boldsymbol{\beta}\gamma$
$E = M\gamma.$
Using the above, one obtains:

$$\beta_{CM} = \frac{\mathbf{P}}{E} \tag{5.21}$$

$$= \frac{(\mathbf{p}_1 + \mathbf{p}_2)}{(E_1 + E_2)}, \tag{5.22}$$

is the velocity of the CM seen from the laboratory system.

$$\gamma_{\text{CM}} = \frac{1}{\sqrt{1 - \beta^2}}$$

$$= \frac{E}{M} \tag{5.23}$$

$$= \frac{E_1 + E_2}{\sqrt{(E_1 + E_2)^2 - (\mathbf{p_1} + \mathbf{p_2})^2}}$$

$$= \frac{E_1 + E_2}{E_{\text{CM}}}, \tag{5.24}$$

is the Lorentz factor or the Lorentz boost of the CM. In general, *the Lorentz factor of the CM is the ratio of the sum of the energies of the particles in the laboratory system and the energy of the CM.*

5.1.4 The Energy, Momentum, and Velocity of One Particle as Seen from the Rest System of Another One

Let us assume that we sit on particle 1, which is in motion. What will be the energy of particle 2, for us?

The answer to this question must always be the same, irrespective of the Lorentz system we start with. It is thus expressible by the invariants discussed in the preceding subsection (point-2), where the last two variables are the well-known Mandelstam's s and t variables.

Let E_{21}: is the energy of particle 2, if we look at it sitting on particle 1, which then appears to be at rest for us.

$E_{21} = E_2$, in the system where $\mathbf{p_1} = 0$. One needs to write E_{21} in terms of the invariants available in this problem.

Now, $p_1 p_2 = E_1 E_2 - \mathbf{p_1} . \mathbf{p_2} = m_1 E_2$ (since $\mathbf{p_1} = 0$).

Hence, $E_{21} = E_2 = \frac{p_1 p_2}{m_1}$.

Note that the RHS is already in an invariant form.

$$|\mathbf{p_{21}}|^2 = E_{21}^2 - m_2^2 = \frac{(p_1 p_2)^2 - m_1^2 m_2^2}{m_1^2}.$$

If p_1 and p_2 are the momentum 4-vectors of any two particles in any Lorentz system, then

$$E_{21} = \frac{p_1 p_2}{m_1}$$

$$|\mathbf{p}_{21}|^2 = \frac{(p_1 p_2)^2 - m_1^2 m_2^2}{m_1^2} \qquad (p_1 p_2 \equiv E_1 E_2 - \mathbf{p}_1 \mathbf{p}_2)$$

$$v_{21}^2 = \frac{|\mathbf{p}_{21}|^2}{E_{21}^2} = \frac{(p_1 p_2)^2 - m_1^2 m_2^2}{(p_1 p_2)^2} \tag{5.25}$$

The above three equations give the energy, momentum, and velocity of particle 2, as seen from particle 1. The velocity v_{21} is the relative velocity, which is symmetric in 1 and 2. Note here that all these expressions are invariant and can be evaluated in any Lorentz system.

5.1.5   The Energy, Momentum, and Velocity of a Particle as Seen from the CM System

This problem is now like all the above quantities are as seen from a *fictitious particle* M, called the "*center-of-momentum-particle*", whose 4-momentum is

$$P = p_1 + p_2 \tag{5.26}$$

We need to apply formulae (Eq. (5.25)) with p_1 replaced by P and p_2 by the 4-momentum of that particle whose energy, momentum, and velocity we would like to determine. From Eq. (5.25),

$$E_1^* = \frac{P p_1}{M}$$

$$|\mathbf{p}_1^*|^2 = \frac{(P p_1)^2 - M^2 m_1^2}{M^2}$$

$$v_1^{*2} = \frac{(P p_1)^2 - M^2 m_1^2}{(P p_1)^2} \tag{5.27}$$

This is by using Eq. (5.26) and by using

$$p_1 p_2 = \frac{1}{2}[(p_1 + p_2)^2 - p_1^2 - p_2^2]$$

$$= \frac{1}{2}(M^2 - m_1^2 - m_2^2), \tag{5.28}$$

one obtains

$$
\begin{aligned}
E_1^* &= \frac{P p_1}{M} \\
&= \frac{(p_1 + p_2) p_1}{M} \\
&= \frac{1}{M}\left[p_1^2 + p_2 p_1 \right] \\
&= \frac{1}{M}\left[p_1^2 + \frac{1}{2}(M^2 - m_1^2 - m_2^2) \right] \\
&= \frac{1}{2M}\left[2m_1^2 + M^2 - m_1^2 - m_2^2 \right] \\
\Rightarrow E_1^* &= \frac{M^2 + (m_1^2 - m_2^2)}{2M}
\end{aligned}
\tag{5.29}
$$

Similarly, one obtains:

$$
\begin{aligned}
E_2^* &= \frac{P p_2}{M} \\
&= \frac{M^2 - (m_1^2 - m_2^2)}{2M}
\end{aligned}
\tag{5.30}
$$

Using the above expressions for E_1^* and E_2^*, one can show that

$$
E_1^* + E_2^* = M
\tag{5.31}
$$

Further,

$$
\begin{aligned}
|\mathbf{p}^*|^2 &= |\mathbf{p_1}^*|^2 = |\mathbf{p_2}^*|^2 \\
&= \frac{M^4 - 2M^2(m_1^2 + m_2^2) + (m_1^2 - m_2^2)^2}{4M^2} \\
&= \frac{[M^2 - (m_1 + m_2)^2][M^2 - (m_1 - m_2)^2]}{4M^2}
\end{aligned}
\tag{5.32}
$$

$$
v_1^{*2} = \left(\frac{|\mathbf{p}^*|}{E_1^*} \right)^2 .
\tag{5.33}
$$

Here, E_1^*, v_1^* are energy and velocity of particle 1, as seen from their common CM system and $M^2 = P^2 = (p_1 + p_2)^2$ is the square of the total mass. The above

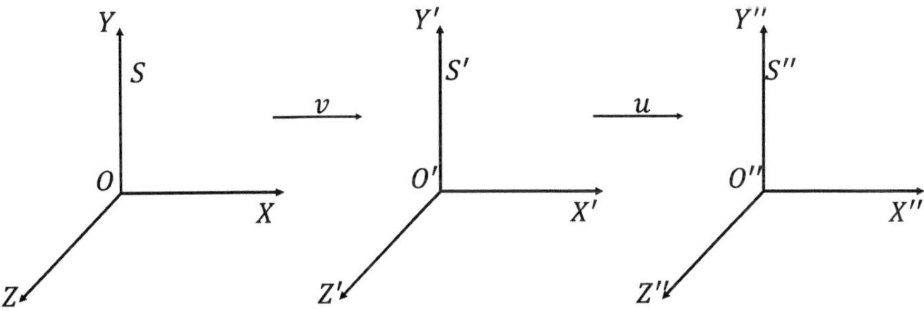

Fig. 5.6 Three inertial frames of reference moving collinearly

equations give the energy, momenta, and velocities of two particles m_1 and m_2 for which

$$M \rightarrow m_1 + m_2$$

5.1.6 What Is the Need for a Variable Called "Rapidity"?

As shown in Fig. 5.6, consider a case where the inertial frame of reference S' moves with a velocity $\mathbf{v} = v\hat{i}$ along the positive X-axis of the inertial frame S. To avoid confusion, let us write the corresponding Lorenz boost factor as:

$$\gamma_v = \frac{1}{\sqrt{1 - \frac{v^2}{c^2}}} \tag{5.34}$$

Omitting other coordinates for simplicity, the corresponding spacetime transformation equations are:

$$x^{0'} = \gamma_v \left(x^0 - \beta_1 x^1 \right)$$

$$x^{1'} = \gamma_v \left(x^1 - \beta_1 x^0 \right)$$

$$\beta_1 = \frac{v}{c} \tag{5.35}$$

Now suppose a third inertial frame of reference S'' is moving at a uniform velocity $\mathbf{u} = u\hat{i}$ with respect to the frame S'. The corresponding Lorenz boost factor is:

$$\gamma_u = \frac{1}{\sqrt{1 - \frac{u^2}{c^2}}} \tag{5.36}$$

The spacetime coordinates in the S'' frame are given by:

$$x^{0''} = \gamma_u \left(x^{0'} - \beta_2 x^{1'} \right)$$

$$x^{1''} = \gamma_u \left(x^{1'} - \beta_2 x^{0'} \right)$$

$$\beta_2 = \frac{u}{c} \tag{5.37}$$

Using Eq. (5.35) in the above equation, we get the relationship of spacetime coordinates in S'' with the coordinates in the frame S by a *single boost*, $\mathbf{w} = w\hat{i}$:

$$x^{0''} = \gamma_w \left(x^0 - \beta x^1 \right)$$

$$x^{1''} = \gamma_w \left(x^1 - \beta x^0 \right)$$

$$\beta = \frac{w}{c}, \tag{5.38}$$

where,

$$\gamma_w = \frac{1}{\sqrt{1 - \frac{w^2}{c^2}}} = \gamma_u \gamma_v \left(1 + \beta_1 \beta_2 \right) = \gamma_u \gamma_v \left(1 + \frac{uv}{c^2} \right) \tag{5.39}$$

and

$$\boxed{\beta = \frac{\beta_1 + \beta_2}{1 + \beta_1 \beta_2}}. \tag{5.40}$$

In terms of usual velocities, the above equation can be written as

$$\boxed{w = \frac{v + u}{1 + \left(vu/c^2 \right)}}. \tag{5.41}$$

Note:

- Successive Lorentz boosts in the same direction are represented by a single boost in the same direction.
- The resulting velocity is **not an additive quantity**, like we get in non-relativistic Newtonian physics. It is **non-linear in successive Lorentz transformations.**
- This non-linearity ensures the ultimate speed limit to the speed of light, c.
- As $|\mathbf{v}| \to c$ or $|\mathbf{u}| \to c$, $|\mathbf{w}| \to c$. This is in line with the postulate of relativity.
- If two Lorentz boosts are in different directions, the resulting boost is equivalent to a single boost together with a rotation, called Wigner rotation.

We have noticed that the formula for the resultant velocity in a successive collinear Lorentz transformation is non-linear in the individual velocities. However, one introduces rapidity, which is a mathematical function of velocity, but follows algebraic addition under a successive collinear Lorentz transformation. *"Rapidity"* is defined as:

$$\boxed{y = \tanh^{-1}(v/c) = \tanh^{-1}\beta}$$
(5.42)

or,

$$y = \frac{1}{2}\ln\left(\frac{1+\beta}{1-\beta}\right)$$
(5.43)

Physically, rapidity is a hyperbolic angle. Let us now show that the rapidity variable is indeed additive under successive Lorentz transformations.

Proof of the Additive Nature of Rapidity via Addition of Velocities
We have already defined the rapidity variable y by

$$\beta \equiv \frac{v}{c} = \tanh y,$$

This gives:

$$\gamma = \cosh y, \qquad \gamma\beta = \sinh y.$$

Following Eq. (5.41), for two collinear boosts with velocities u and v, the relativistic velocity addition formula with $c = 1$, gives

$$w = \frac{u+v}{1+uv}.$$

Now substituting $u = \tanh y_1$ and $v = \tanh y_2$, we get

$$w = \frac{\tanh y_1 + \tanh y_2}{1 + \tanh y_1 \, \tanh y_2}.$$

Using the hyperbolic tangent addition identity,

$$\tanh(y_1 + y_2) = \frac{\tanh y_1 + \tanh y_2}{1 + \tanh y_1 \, \tanh y_2},$$

we get

$$w = \tanh(y_1 + y_2).$$

By definition, w is the resulting velocity and hence, $w = \tanh y_{\text{tot}}$. Therefore,

$$\boxed{y_{\text{tot}} = y_1 + y_2}.$$

This additive nature of rapidity can also be shown by taking the boost matrices, which we leave to the readers as an exercise.

In the subsequent sections, when we plan to discuss the rapidity variable in detail, we shall show that the rapidity distribution follows a shape invariance under a Lorentz transformation. This has an added advantage while drawing physics information in case of a transformation from the center-of-mass system to the laboratory system and vice versa.

5.2   Description of Nucleus-Nucleus Collisions in terms of Light-Cone Variables

In relativistic nucleus-nucleus collisions, it is convenient to use kinematic variables that take simple forms under Lorentz transformations for the change of frame of reference. A few of them are the light cone variables x_+ and x_-, the rapidity and pseudorapidity variables, y and η. A particle is characterized by its 4-momentum, $p_\mu = (E, \mathbf{p})$. In fixed target and collider experiments where the beam(s) define reference frames, boosted along their direction, it is important to express the 4-momentum in terms of more practical kinematic variables.

Figure 5.7 shows the collision of two Lorentz contracted nuclei approaching each other with velocities nearly equal to the velocity of light. The vertical axis represents the time direction, with the lower half representing time before the collision and the upper half, time after the collision. The horizontal axis represents the spatial direction. Both the nuclei collide at $(t, z) = (0, 0)$ and then the created fireball expands in time, going through various processes till the created particles freeze-out and reach the detectors. The lines where $t^2 - z^2 = 0$ (note that $\sqrt{t^2 - z^2} \equiv \tau$, τ being the proper time of the particle) along the path of the colliding nuclei defines the light cone. The upper part of the light-cone, where $t^2 - z^2 > 0$, is the time-like region. In nucleus-nucleus collisions, particle production occurs in the upper half of the (t, z)-plane within the light-cone. The region outside the light cone for which $t^2 - z^2 < 0$ is called a space-like region. The space-time rapidity is defined as

$$\eta_s = \frac{1}{2} ln \left(\frac{t + z}{t - z} \right) \tag{5.44}$$

It could be seen that η_s is not defined in the space-like region. It takes the value of positive and negative infinity along the beam directions for which $t = \pm z$ respectively. A particle is "light-like" along the beam direction. Inside the light-cone, which is time-like, η_s is properly defined.

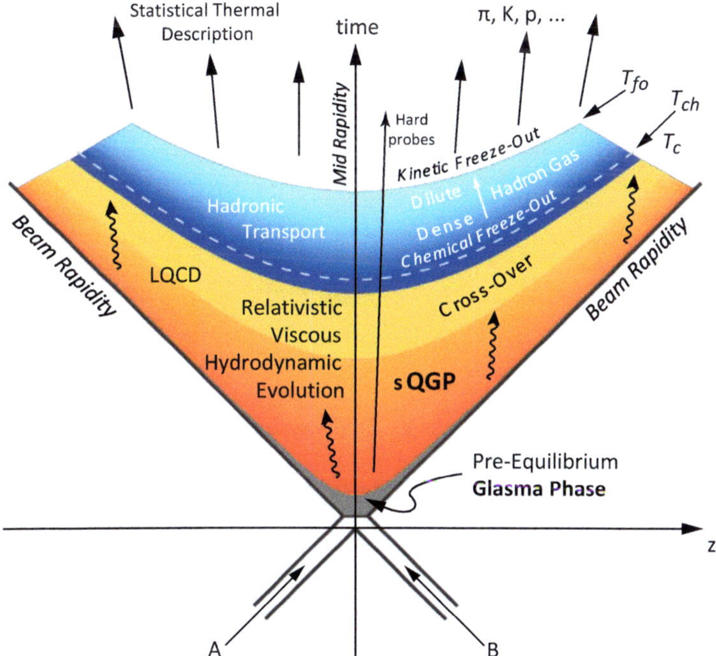

Fig. 5.7 Description of heavy-ion collisions in one space (z) and one time (t) dimension, with two nuclei colliding at ($t = 0, z = 0$). The formation of a very hot and dense fireball is envisaged, which follows a complex and dynamical spacetime evolution involving different quanta of the system at different time scales. The fireball passes through a pre-equilibrium phase, followed by the formation of a QGP and a cross-over phase transition to a hadron gas. This transition is marked by a critical temperature, T_c after which the chemical freeze-out and kinetic freeze-out of the system happen at temperatures, T_{ch} and T_{kin}, respectively. Final state secondary particles free-stream after the kinetic freeze-out to reach the detectors. Image courtesy of Prof. Boris Hyppolyte. All rights reserved

For a particle with 4-momentum $p = (p_0, \mathbf{p}) = (p_0, \mathbf{p_T}, p_z)$, the light-cone momenta are defined by

$$p_+ = p_0 + p_z \tag{5.45}$$

$$p_- = p_0 - p_z \tag{5.46}$$

p_+ is called "*forward light-cone momentum*" and p_- is called "*backward light-cone momentum*".

For a particle travelling along the beam direction, has a higher value of forward light-cone momentum and travelling opposite to the beam direction has a lower value of forward light-cone momentum. The advantages of using light-cone variables to study particle production are the following.

1. The forward light-cone momentum of any particle in one frame is related to the forward light-cone momentum of the same particle in another boosted Lorentz frame by a constant factor.
2. Hence, if a daughter particle c is fragmenting from a parent particle b, then the ratio of the forward light-cone momentum of c relative to that of b is independent of the Lorentz frame.

Define

$$x_+ = \frac{p_0^c + p_z^c}{p_0^b + p_z^b} \tag{5.47}$$

$$= \frac{c_+}{b_+}.$$

The forward light-cone variable x_+ is always positive. The upper limit of x_+ is 1, because c_+ can't be greater than b_+. x_+ is Lorentz invariant.

3. The Lorentz invariance of x_+ provides a tool to measure the momentum of any particle in the scale of the momentum of any reference particle.

5.2.1   Pictorial Representation of Detector System

In a collider environment, usually, the collision axis is chosen along the z-direction (longitudinal), and hence, the $(x - y)$-plane is the transverse plane. A particle is emitted from the collision point, the global $(0, 0, 0)$ making a polar angle θ with the collision point. When the momenta of the particles are determined by a tracking detector, the (pseudo)-rapidity and azimuthal angles are given by:

$$\beta = \tanh y$$

$$\Rightarrow y = \tanh^{-1} \beta = \tanh^{-1} \left(\frac{|\mathbf{p}|}{E} \right). \tag{5.48}$$

If the velocity is along the longitudinal direction,

$$y = \tanh^{-1} v_z$$

$$= \tanh^{-1} \frac{p_z}{E}$$

$$= \tanh^{-1} \frac{p_z}{\sqrt{p_x^2 + p_y^2 + p_z^2 + m_0^2}}, \tag{5.49}$$

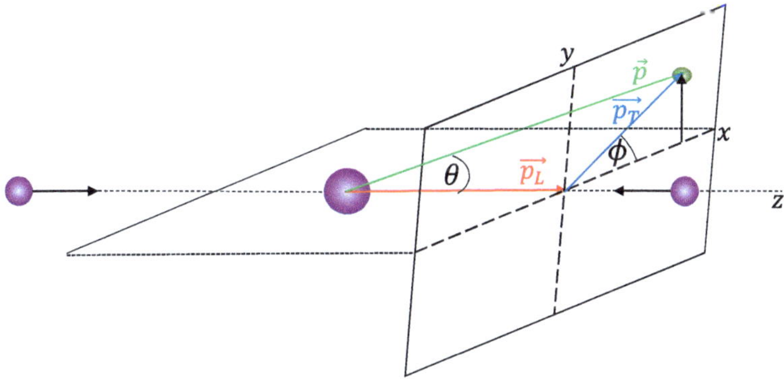

Fig. 5.8 A schematic decomposition of particle momentum **p** (in CM frame) into longitudinal and transverse components. Note the angle of inclination θ of **p** and the azimuthal angle ϕ of p_T

where, m_0 is the rest mass of the particle, and

$$\phi = tan^{-1}\frac{p_y}{p_x}. \tag{5.50}$$

The polar angle θ is given by:

$$\theta = \cos^{-1}\frac{p_z}{|\mathbf{p}|}$$
$$= \tan^{-1}\frac{|\mathbf{p_T}|}{p_z}. \tag{5.51}$$

Pictorially, these are shown in the Fig. 5.8. A detector plane is spanned by (η, ϕ), with η decreasing while going away from the beam axis in annular rings and ϕ is scanned making an angle with the beam axis and increasing it anti-clockwise, as is shown in the picture.

5.2.2 The Rapidity Variable

One of the important tasks at hand is to introduce key kinematic variables that relate particle momentum to the dynamics that are occurring in heavy-ion collisions. It is essentially the convenience of working in a center-of-momentum system; we need to introduce the observable rapidity. To do so, let's proceed as follows: The coordinates along the beam line (conventionally along the z-axis) are called *longitudinal* and perpendicular to it is called *transverse* (x-y). The 3 momentum can be decomposed into the longitudinal (p_z) and the transverse ($\mathbf{p_T}$), $\mathbf{p_T}$ being a vector quantity which is invariant under a Lorentz boost along the longitudinal direction. The variable rapidity "y" is defined by

$$y = \frac{1}{2} \ln \left(\frac{E + p_z}{E - p_z} \right) \tag{5.52}$$

$$= \frac{1}{2} \ln \left(\frac{1 + \beta}{1 - \beta} \right) \tag{5.53}$$

$$= \frac{1}{2} \ln \left(\frac{(E + p_z)^2}{m^2 + p_T^2} \right)$$

$$= \ln \left(\frac{(E + p_z)}{\sqrt{m^2 + p_T^2}} \right)$$

$$= \ln \left(\frac{E + p_z}{m_T} \right) \tag{5.54}$$

It is a dimensionless quantity related to the ratio of forward light-cone to backward light-cone momentum. The rapidity changes by an additive constant under longitudinal Lorentz boosts.

Note that rapidity can also be treated as an angle of rotation under the Lorentz transformation:

$$\begin{bmatrix} p_z^* \\ E^* \end{bmatrix} = \begin{bmatrix} \gamma & -\gamma\beta \\ -\gamma\beta & \gamma \end{bmatrix} \begin{bmatrix} p_z \\ E \end{bmatrix}$$

In analogy with the orthogonal transformation:

$$\begin{bmatrix} x_1' \\ x_2' \end{bmatrix} = \begin{bmatrix} \cos\phi & -\sin\phi \\ \sin\phi & \cos\phi \end{bmatrix} \begin{bmatrix} x_1 \\ x_2 \end{bmatrix},$$

let us define an angle y of transformation such that:

$$\cosh y = \gamma \equiv \frac{E}{m} \tag{5.55}$$

$$\sinh y = \gamma\beta \equiv \frac{p_z}{m} \tag{5.56}$$

Recall:

$$\cosh y = (e^y + e^{-y})/2, \tag{5.57}$$

$$\sinh y = (e^y - e^{-y})/2, \tag{5.58}$$

$$\cosh^2 y - \sinh^2 y = 1. \tag{5.59}$$

Taking these into account, the above Lorentz transformation becomes:

$$\begin{bmatrix} p_z^* \\ E^* \end{bmatrix} = \begin{bmatrix} \cosh y & -\sinh y \\ -\sinh y & \cosh y \end{bmatrix} \begin{bmatrix} p_z \\ E \end{bmatrix}$$

One can see that, as expected for an orthogonal transformation, the determinant of the transformation matrix is unity. As can be seen from Eqs. (5.55) and (5.56),

$$y = \tanh^{-1}(\beta), \tag{5.60}$$

where β is the velocity along the longitudinal axis, i.e, the beam axis.

For a free particle which is on the mass shell (for which $E^2 = \mathbf{p}^2 + m_0^2$), the 4-momentum has only three degrees of freedom and can be represented by $(y, \mathbf{p}_T)$. $(E, \mathbf{p}_T)$ could be expressed in terms of $(y, \mathbf{p}_T)$ as

$$E = m_T \cosh y \tag{5.61}$$

$$p_z = m_T \sinh y \tag{5.62}$$

m_T being the transverse mass, which is defined as:

$$m_T^2 = m_0^2 + \mathbf{p}_T^2. \tag{5.63}$$

The advantage of the rapidity variable is that the shape of the rapidity distribution remains unchanged under a longitudinal Lorentz boost. When we go from CMS to LS, the rapidity distribution is the same, with the y-scale shifted by an amount equal to y_{cm}. This is shown below.

5.2.2.1 Beam and Target Rapidities in the Laboratory System

Following the above definition of rapidity (Eq. (5.54)), the beam rapidity in the laboratory system can be written as:

$$y_{lab}^{beam} = \frac{1}{2} \ln \left(\frac{E_{lab} + p_{lab}}{E_{lab} - p_{lab}} \right) \tag{5.64}$$

$$= \ln \left[\gamma (1 + \beta) \right] \tag{5.65}$$

Here, we have used $p_z = p_{lab}$, as the beam is along the longitudinal direction. For large E_{lab}, $\beta \simeq 1$, so that y_{lab}^{beam} becomes:

$$y_{lab}^{beam} \simeq \ln(2\gamma) \tag{5.66}$$

$$- \ln \left(\frac{2E_{lab}}{m_a} \right) \tag{5.67}$$

$$= \ln\left(\frac{s}{m_a m_b}\right) \tag{5.68}$$

$$\Rightarrow s \simeq m_a m_b e^{y_{beam}}. \tag{5.69}$$

Following Eq. (5.9), here, one uses $s = 2m_b E_{lab}$, with m_a and m_b being the projectile and target masses, respectively.

Target rapidity in the laboratory system is: $y_{lab}^{target} = 0$, as the target is at rest.

5.2.2.2 Rapidity of Center-of-Mass in the Laboratory System

The total energy in the CMS system is:

$$E_{cm} = \sqrt{s}.$$

The energy and longitudinal momentum of the CMS in the LS are $\gamma_{cm}\sqrt{s}$ and $\beta_{cm}\gamma_{cm}\sqrt{s}$, respectively. This is because, $\gamma_{cm} = \frac{E_{lab}}{\sqrt{s}}$. The rapidity of the CMS in the LS is

$$\begin{aligned}
y_{cm} &= \frac{1}{2}\ln\left[\frac{\gamma_{cm}\sqrt{s} + \beta_{cm}\gamma_{cm}\sqrt{s}}{\gamma_{cm}\sqrt{s} - \beta_{cm}\gamma_{cm}\sqrt{s}}\right] \\
&= \frac{1}{2}\ln\left[\frac{1 + \beta_{cm}}{1 - \beta_{cm}}\right]
\end{aligned} \tag{5.70}$$

It is a constant for a given Lorentz boost.

5.2.2.3 Relationship Between Rapidity of a Particle in LS and Rapidity in CMS

The rapidities of a particle in the LS and CMS of the collision are respectively,

$$y = \frac{1}{2}\ln\left(\frac{E + p_z}{E - p_z}\right), \tag{5.71}$$

and

$$y^* = \frac{1}{2}\ln\left(\frac{E^* + p_z^*}{E^* - p_z^*}\right). \tag{5.72}$$

For a particle travelling in the longitudinal direction, the Lorentz transformation of its energy and momentum components gives

$$\begin{bmatrix} E^* \\ p_L^* \end{bmatrix} = \begin{bmatrix} \gamma & -\gamma\beta \\ -\gamma\beta & \gamma \end{bmatrix} \cdot \begin{bmatrix} E \\ p_L \end{bmatrix}, \quad p_T^* = p_T \tag{5.73}$$

where p_L and p_T are the longitudinal and transverse components of $\mathbf{p}$, which are parallel and perpendicular to β, respectively. Hence, the inverse Lorentz transformations on E and p_z give:

$$y = \frac{1}{2} \ln \left[\frac{\gamma(E^* + \beta p_z^*) + \gamma(\beta E^* + p_z^*)}{\gamma(E^* + \beta p_z^*) - \gamma(\beta E^* + p_z^*)} \right]$$

$$= \frac{1}{2} \ln \left[\frac{E^* + p_z^*}{E^* - p_z^*} \right] + \frac{1}{2} \ln \left[\frac{1+\beta}{1-\beta} \right] \tag{5.74}$$

$$\Rightarrow y = y^* + y_{cm}. \tag{5.75}$$

Hence, the rapidity of a particle in the laboratory system (y) is equal to the sum of the rapidity of the particle in the center-of-mass system (y^*) and the rapidity of the center-of-mass in the laboratory system (y_{cm}). It can also be stated that the rapidity of a particle in a moving (boosted) frame is equal to the rapidity in its own rest frame minus the rapidity of the moving frame. In the non-relativistic limit, this is like the subtraction of the velocity of the moving frame. However, this is not surprising because, non-relativistically, the rapidity y is equal to longitudinal velocity β. Rapidity is a relativistic measure of velocity. This simple property of the rapidity variable under the Lorentz transformation makes it a suitable choice to describe the dynamics of relativistic particles. The simple shape invariance nature of rapidity spectra brings its importance in the analysis of particle production in nuclear collisions. For instance, in fixed-target experiments, we can study particle spectra using y as a variable without making an explicit transformation to the CM frame of reference, and from the rapidity spectra, we deduce the point of symmetry corresponding to the CM rapidity. In symmetric collisions with fixed targets, the CM frame is located in the middle between the rapidities of the projectile and target i.e. $y_{cm} = y_{proj}/2$. In this case, the particle rapidity spectrum must be symmetric around y_{cm}. This allows for complementing measured particle spectra: if these are available for, e.g. $y \geq y_{cm}$, a reflection at the symmetry point y_{cm} gives us the part of the spectrum with $y \leq y_{cm}$, for which an experimental measurement is absent.

5.2.2.4 Relationship Between Rapidity and Velocity

Consider a particle travelling in the z-direction with a longitudinal velocity β. The energy E and the longitudinal momentum p_z of the particle are:

$$E = \gamma m_0 \tag{5.76}$$

$$p_z = \gamma \beta m_0 \tag{5.77}$$

$$p_z = \beta E, \tag{5.78}$$

where m_0 is the rest mass of the particle. Hence, the rapidity of the particle travelling in z-direction with velocity β is:

$$y_\beta = \frac{1}{2} \ln \left[\frac{E + p_z}{E - p_z} \right] = \frac{1}{2} \ln \left[\frac{\gamma m_0 + \gamma \beta m_0}{\gamma m_0 - \gamma \beta m_0} \right]$$

$$= \frac{1}{2} \ln \left[\frac{1 + \beta}{1 - \beta} \right] \tag{5.79}$$

Note here that y_β *is independent of particle mass*. A particle is said to be relativistic in nature if $\gamma >> 1$, $\beta \approx 1$ and $E \approx p$. In other words, the energy of the particle is much higher than its rest mass i.e. $E >> m_0$ and hence the energy and the total momentum of the particle is comparable. In the non-relativistic limit when β is small, expanding y_β in terms of β leads to

$$y_\beta = \beta + O(\beta^3) \tag{5.80}$$

Thus, the rapidity of the particle is the relativistic realization of its velocity.

5.2.2.5 Beam Rapidity

We know,

$$E = m_T \, \cosh y,$$

$$p_z = m_T \, \sinh y, \tag{5.81}$$

$$m_T^2 = m_0^2 + \mathbf{p}_T^2.$$

For the beam particles, $p_T = 0$, as the beam is along the longitudinal direction. Hence,

$$E = m_b \, \cosh y_b$$

$$p_z = m_b \, \sinh y_b, \tag{5.82}$$

where m_b and y_b are the rest mass and rapidity of the beam particles. From Eq. (5.82), we get:

$$\boxed{y_b = \cosh^{-1} (E/m_b)} \tag{5.83}$$

Note that Eq. (5.83) is used for a fixed target experiment with E as the energy of individual beam particles. For example, for SPS beam energy of 158 AGeV, $E = 158$ GeV. This leads to $y_b = 5.82$ in the fixed target framework. Now, when we convert Eq. (5.83) for a collider system, we get

$$y_b = \cosh^{-1} (E/m_b)$$

$$= \cosh^{-1} \left[\frac{\sqrt{s_{NN}}}{2\,m_p} \right] \qquad (5.84)$$

Here, m_p is the mass of a proton. We know:

$$\cosh^{-1} x = \ln(x + \sqrt{x^2 - 1}) \simeq \ln(2x)$$

$$= \ln \left(2 \cdot \frac{\sqrt{s_{NN}}}{2\,m_p} \right) \qquad (5.85)$$

$$= \ln \left(\frac{\sqrt{s_{NN}}}{m_p} \right).$$

This approximation is valid for $x \gg 1$. Here, $x = \frac{\sqrt{s_{NN}}}{2m_p} \gg 1$. Using this, one obtains:

$$\boxed{ y_b \simeq \ln \left(\frac{\sqrt{s_{NN}}}{m_p} \right) } \qquad (5.86)$$

Now, for the above SPS beam energy of 158 AGeV, the equivalent center of mass energy per nucleon is $\sqrt{s_{NN}} = 17.3\,\text{GeV}$. By using Eq. (5.86), we get $y_b = 2.91$. This makes sense to us, as in a collider experiment, both the beams come in opposite directions with equal beam rapidities.

Further,

$$y_b = \sinh^{-1} (p_z / m_b) \qquad (5.87)$$

We know,

$$\frac{E}{p_z} = \frac{m_b \cosh y_b}{m_b \sinh y_b}$$

$$\Rightarrow \tanh y_b = \frac{p_z}{E}.$$

In terms of the beam velocity, beam rapidity can be expressed as follows:

$$\boxed{ y_b = \tanh^{-1} \left(\frac{p_z}{E} \right) = \tanh^{-1} \beta } \qquad (5.88)$$

Note that the beam energy $E = \frac{\sqrt{s_{NN}}}{2}$ for a symmetric collider.

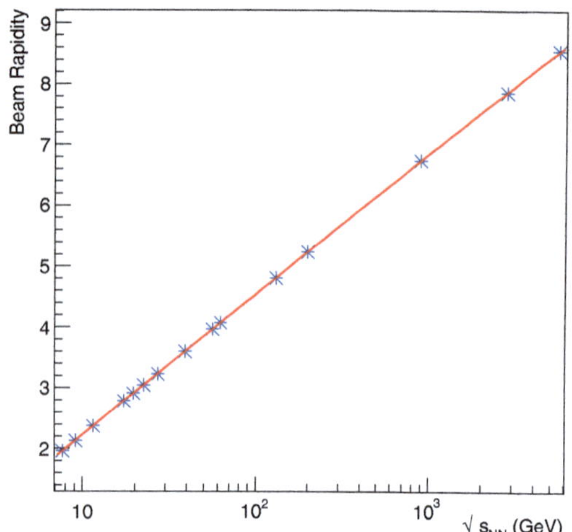

Fig. 5.9 A plot of beam rapidity as a function of center-of-mass energy

Table 5.2 Table of collision energy, $\sqrt{s_{NN}}$ vs the beam rapidity, y_b

$\sqrt{s_{NN}}$ (GeV)	7.7	9.1	11.5	17.3	19.6	27.0	39.0	62.4
y_b	2.10	2.27	2.50	2.91	3.03	3.35	3.72	4.19
$\sqrt{s_{NN}}$ (GeV)	130	200	900	2360	2760	5520	7000	14,000
y_b	4.93	5.36	6.86	7.83	7.98	8.68	8.91	9.61

Furthermore, it could be shown that:

$$y_b \simeq \mp \ln\left(\frac{\sqrt{s_{NN}}}{m_p}\right) \simeq \mp y_{\max}. \tag{5.89}$$

Figure 5.9 shows the variation of beam rapidity as a function of the collision center-of-mass energy in a semi-log plot. Table 5.2 enlists the estimated values of beam rapidities for a set of center-of-mass energies.

Example 5.3 Beam rapidity

Solution 5.3 For the nucleon-nucleon center-of-mass energy $\sqrt{s_{NN}} = 9.1$ GeV, the beam rapidity $y_b = \cosh^{-1}\left(\frac{9.1}{2 \times 0.938}\right) = 2.26$

For p+p collisions with lab momentum 100 GeV/c, the beam rapidity

$$y_b = \sinh^{-1}\left(\frac{p_z}{m_b}\right) = \sinh^{-1}\left(\frac{100}{0.938}\right) = 5.36$$

and for Pb+Pb collisions at SPS with lab energy 158 AGeV, $y_b = 2.91$.

5.2.2.6 Rapidity of the CMS in Terms of Projectile and Target Rapidities

Let us consider the beam particle "b" and the target particle "a". The longitudinal momentum of the beam particle,

$$b_z = m_T \ \sinh y_b = m_b \ \sinh y_b. \tag{5.90}$$

This is because p_T of beam particles is zero. Hence,

$$y_b = \ \sinh^{-1} \ (b_z/m_b). \tag{5.91}$$

The energy of the beam particle (b_0) in the laboratory frame is:

$$b_0 = \ m_T \ \cosh \ y_b = \ m_b \ \cosh \ y_b. \tag{5.92}$$

Assuming target particle a has longitudinal momentum a_z, its rapidity in the laboratory frame is given by

$$y_a = \ \sinh^{-1} \ (a_z/m_a) \tag{5.93}$$

and its energy

$$a_0 = \ m_a \ \cosh \ y_a. \tag{5.94}$$

The CMS is obtained by boosting the LS by the velocity of the center-of-mass frame β_{cm} such that the longitudinal momentum of the beam particle b_z^* and of the target particle a_z^* are equal and opposite. Hence β_{cm} satisfies the condition,

$$a_z^* = \gamma_{cm}(a_z - \beta_{cm}a_0) = -b_z^*$$
$$= -\gamma_{cm}(b_z - \beta_{cm}b_0), \tag{5.95}$$

where $\gamma_{cm} = \dfrac{1}{\sqrt{1-\beta_{cm}^2}}$.

Hence,

$$\beta_{cm} = \frac{a_z + b_z}{a_0 + b_0}. \tag{5.96}$$

We know the rapidity of the center-of-mass is

$$y_{cm} = \frac{1}{2} \ \ln \left[\frac{1+\beta_{cm}}{1-\beta_{cm}} \right] \tag{5.97}$$

Using Eqs. (5.96) and (5.97), we get

$$y_{cm} = \frac{1}{2} \ln \left[\frac{a_0 + a_z + b_0 + b_z}{a_0 - a_z + b_0 - b_z} \right]. \tag{5.98}$$

Writing energies and momenta in terms of rapidity variables in the LS,

$$y_{cm} = \frac{1}{2} \ln \left[\frac{m_a \cosh y_a + m_a \sinh y_a + m_b \cosh y_b + m_b \sinh y_b}{m_a \cosh y_a - m_a \sinh y_a + m_b \cosh y_b - m_b \sinh y_b} \right]$$

$$= \frac{1}{2}(y_a + y_b) + \frac{1}{2} \ln \left[\frac{m_a \, e^{y_a} + m_b \, e^{y_b}}{m_a \, e^{y_b} + m_b \, e^{y_a}} \right] \tag{5.99}$$

For a symmetric collision (for $m_a = m_b$),

$$y_{cm} = \frac{1}{2}(y_a + y_b) \tag{5.100}$$

Rapidities of a and b in the CMS are

$$y_a^* = y_a - y_{cm} = -\frac{1}{2}(y_b - y_a) \tag{5.101}$$

$$y_b^* = y_b - y_{cm} = \frac{1}{2}(y_b - y_a). \tag{5.102}$$

Given the incident energy or the longitudinal momentum, or the lab momentum, the rapidity of projectile particles, and the rapidity of the target particles can thus be determined. The greater the incident energy, the greater is the separation between the projectile and target rapidity.

Central Rapidity

The region of rapidity midway between the projectile and the target rapidities is called central rapidity.

$$y_{cm} = \frac{1}{2}(y_a + y_b)$$

Example 5.4   Beam rapidity, target rapidity, and central rapidity

Solution 5.4 In p+p collisions at a laboratory momentum of 250 GeV/c, beam rapidity, $y_b = \sinh^{-1}(\frac{b_z}{m_b}) \simeq \sinh^{-1}(250) \simeq 6.2$, target rapidity, $y_a = 0$, and the central rapidity ≈ 3.1.

5.2.2.7 Mid-rapidity in Fixed Target and Collider Experiments

In fixed-target experiments (LS), $y_{target} = 0$.

$$y_{lab} = y_{target} + y_{projectile} = y_{beam}$$

Hence, mid-rapidity in a fixed-target experiment is given by,

$$y_{mid}^{LS} = y_{beam}/2. \tag{5.103}$$

In collider experiments (center-of-mass system),

$$y_{projectile} = -y_{target} = y_{CMS} = y_{beam}/2.$$

Hence, mid-rapidity in the CMS system is given by

$$y_{mid}^{CMS} = (y_{projectile} + y_{target})/2 = 0. \tag{5.104}$$

This is valid for a symmetric energy collider. The rapidity difference is given by $y_{projectile} - y_{target} = 2y_{CMS}$ and this increases with energy for a collider as y increases with energy.

5.2.2.8 Light-cone Variables and Rapidity

Consider a particle c, having rest mass m_c, that is detected to have rapidity y in a given coordinate system. The beam rapidity of the beam particle, b is y_b. Given m_b is the rest mass of the beam particle. The detected particle has a forward light cone variable x_+. The forward light-cone variable x_+ is related to the masses and the rapidities as:

$$x_+ = \frac{p_{c0} + p_{cz}}{p_{b0} + p_{bz}}$$

$$= \frac{m_{cT}}{m_b} e^{y - y_b} \tag{5.105}$$

where m_{cT} is the transverse mass of c. Note that the transverse momentum of the beam particle is zero. Hence,

$$y = y_b + \ln x_+ + \ln\left(\frac{m_b}{m_{cT}}\right) \tag{5.106}$$

Similarly, relative to the target particle a with a target rapidity y_a, the backward light-cone variable of the detected particle c is x_-. x_- is related to y by

$$x_- = \frac{m_{cT}}{m_b} e^{y_a - y} \tag{5.107}$$

and conversely,

$$y = y_a - \ln x_- - \ln\left(\frac{m_a}{m_{cT}}\right).$$

(5.108)

In general, the rapidity of a particle is related to its light-cone momenta by

$$y = \frac{1}{2}\ln\left(\frac{p_+}{p_-}\right)$$

(5.109)

Note that in situations where there is a frequent need to work with boosts along z-direction, it is better to use $(y, \mathbf{p}_T)$ for a particle rather than using its 3-momentum, because of the simple transformation rules for y and $\mathbf{p}_T$ under Lorentz boosts.

5.2.2.9 The Maximum Accessible Rapidity in an Interaction
We know the variable rapidity "y" of a particle is defined by Eq. (5.54):

$$y = \frac{1}{2}\ln\left(\frac{E + p_z}{E - p_z}\right).$$

(5.110)

This corresponds to

$$\tanh y = \frac{p_z}{E},$$

(5.111)

where p_z is the longitudinal momentum along the direction of the incident particle, E is the energy, both defined for a given particle. The accessible range of rapidities for a given interaction is determined by the available centre-of-mass energy and all participating particles' rest masses. One usually gives the limit for the incident particle, elastically scattered at zero angle:

$$
\begin{aligned}
|y|_{max} &= \ln[(E + p_z)/m] \\
&= \ln[\gamma + \gamma\beta] \\
&= \ln[\gamma + \sqrt{\gamma^2 - 1}] \\
&\simeq \cosh^{-1}\gamma, \quad if \ \gamma \gg 1
\end{aligned}
$$

(5.112)

with all variables referring to the through-going particle given in the desired frame of reference (e.g., in the centre of mass). A Lorentz boost β along the direction of the incident particle adds a constant, $\ln[\gamma + \gamma\beta]$, to the rapidity. Rapidity differences, therefore, are invariant to a Lorentz boost. Statistical particle distributions are flat in y for many physics production models. Note here that $\frac{\partial y}{\partial p_z} = 1/E$.

Furthermore,

$$\gamma = \frac{E_{beam}}{m_p} \tag{5.113}$$

and for a symmetric collision,

$$E_{beam} = \frac{\sqrt{s_{NN}}}{2} \Rightarrow \gamma = \frac{\sqrt{s_{NN}}}{2m_p} \tag{5.114}$$

Hence,

$$y_{max} = cosh^{-1}\left[\frac{\sqrt{s_{NN}}}{2m_p}\right]$$

$$\Rightarrow \boxed{y_{max} = y_b = ln\left[\frac{\sqrt{s_{NN}}}{m_p}\right]} \tag{5.115}$$

Note that the maximum accessible rapidity is independent of the collision species for a symmetric collider and only depends on the center-of-mass energy. The center-of-mass energy of various accelerators as a function of the projectile and target rapidities seen from the CM frame is shown in Fig. 5.10. The shaded areas depict energy ranges accessible at various accelerators.

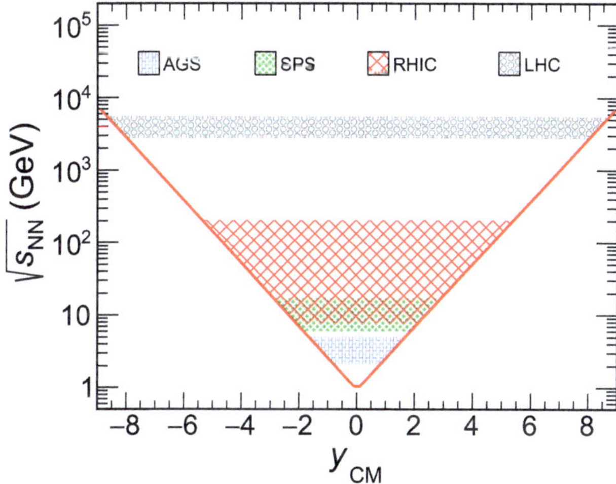

Fig. 5.10 The center-of-mass-energy, $\sqrt{s_{NN}}$ (vertical axis) of various accelerators as a function of the projectile and target rapidities seen from the CM frame. Shaded areas; energy ranges accessible at various accelerators

Example 5.5 Maximum rapidity

Solution 5.5

(a) For RHIC top energy, $\sqrt{s_{NN}} = 200\,\text{GeV}$, $\gamma = E_{beam}/m_p = 106.609$. Hence $y_{max} = 5.36$. For LHC, center-of-mass energy of $5.5\,\text{TeV}$ ($\gamma = 2931.768$), $y_{max} = 8.67$.

(b) For a fixed-target setup, $y_{beam}^{lab} = 5.8$ for $E_{lab} = 158\,\text{AGeV}$, and $y_{beam}^{lab} = 4.4$ for $E_{lab} = 40\,\text{AGeV}$.

Example 5.6 Range of Rapidity for Large CM Energy

Consider an interaction: $a + b \rightarrow c + X$, where "a" is projectile and "b" is the target. Find the rapidity range for this interaction.

Solution 5.6 The transverse momentum for the projectile and target are: $p_{aT} = p_{bT} = 0$. In the CM system, for large collision energy (large value of $\sqrt{s}$), $p_a^* = -p_b^* = p^* \approx \sqrt{s}/2$.

$$y_a^* = \ln\left(\frac{E_a^* + p_a^*}{\sqrt{m_a^2 + p_{aT}^2}}\right) \tag{5.116}$$

$$\simeq \ln\left(\frac{E_a^* + p_a^*}{m_a}\right)$$

$$\simeq \ln\left(\frac{\sqrt{s}}{m_a}\right)$$

Similarly, one can write,

$$y_b^* = \frac{1}{2}\ln\left(\frac{E_b^* + p_b^*}{E_b^* - p_b^*}\right) \tag{5.117}$$

$$= \frac{1}{2}\ln\left(\frac{E_b^* - p^*}{E_b^* + p^*}\right)$$

$$= -\ln\left(\frac{E_b^* + p^*}{\sqrt{m_b^2 + p_{bT}^2}}\right)$$

$$\simeq -\ln\left(\frac{E_b^* + p^*}{m_b}\right)$$

$$\simeq -\ln\left(\frac{\sqrt{s}}{m_b}\right)$$

The rapidity gap between the beam particles, 'a' and 'b' in the CM system is:

$$y_a^* - y_b^* \simeq \ln\left(\frac{\sqrt{s}}{m_a}\right) + \ln\left(\frac{\sqrt{s}}{m_b}\right) \tag{5.118}$$

$$= \ln\left(\frac{s}{m_a m_b}\right) \tag{5.119}$$

This means the rapidity range goes like $\ln s$, and the produced secondary particles will have rapidities in this interval. In the laboratory system:

$$y_b = 0 \tag{5.120}$$

$$y_a = \ln\left(\frac{E_a^{lab} + p_a^{lab}}{\sqrt{m_a^2 + p_{aT}^2}}\right)$$

$$\simeq \ln\left(\frac{E_a^{lab} + p_a^{lab}}{m_a}\right)$$

$$\simeq \ln\left(\frac{2E_a^{lab}}{m_a}\right)$$

$$\simeq \ln\left(\frac{s}{m_a m_b}\right) \tag{5.121}$$

$s \simeq 2m_b E_a^{lab}$ is used in the above equation. *Note that the rapidity gap becomes independent of the frame of reference.*

5.2.2.10 Notes on Properties of Rapidity
1. Rapidity is the relativistic realization of particle velocity.
2. Rapidity is a dimensionless quantity, which describes the rate at which a particle moves with respect to a chosen reference point situated on the trajectory of motion.
3. Rapidity,

$$y = \tanh^{-1}\beta = \frac{1}{2}\ln\left[\frac{1+\beta}{1-\beta}\right] \tag{5.122}$$

4. In terms of the energy, E and momentum, p of the particle,

$$y = \tanh^{-1}\beta$$

$$= \tanh^{-1}(p/E)$$

$$= \tanh^{-1} \left(\frac{\gamma \beta m}{\gamma m} \right)$$

$$= \frac{1}{2} \ln \left(\frac{E + p}{E - p} \right) \tag{5.123}$$

It is difficult/not always possible to measure E and/or p in an experiment, as energy measurement requires calorimetry or, with trackers, the precise identification of secondary particles (PID) with momentum information, in order to estimate the rapidity of a particle. To circumvent this problem, one introduces a variable called *psuedorapidity* ($\equiv \eta = -\ln[\tan(\theta/2)]$), which only depends on the measurement of the polar angle of emission, θ, i.e., the angle a particle makes with the detector plane with respect to the collision axis. This will be discussed in detail in the subsequent sections.

5. The notion of positive and negative rapidity is purely a convention. The positive or negative rapidity corresponds to positive or negative velocities of a particle with respect to a chosen axis.

6. As rapidity has a logarithmic dependence on energy and momentum of a particle, the magnitude of rapidity is rather small, even for the most energetic particle.

7. The range of rapidity in principle is: $-\infty \leq y \leq +\infty$.

8. In a collider geometry, the rapidity variable makes annular rings on the detector plane, having higher values towards the beam pipe. Both rapidity, y, and the azimuthal angle, ϕ, span the detector plane.

9. It can be shown explicitly that the *rapidity formulation simply and naturally incorporates the Lorentz transformation properties of velocity*. This will be clear with the example, which follows these notes.

10. The differences in rapidities are Lorentz Invariants. This is also true for the differential element of rapidity, dy, which will soon be seen as important in understanding the invariant cross-section.

11. *Shape invariance of rapidity distribution under a Lorentz transformation from the CM frame to the LS frame or vice versa is an important property for which rapidity is a chosen variable.* The rapidity of a particle in the LS frame (y) is related to the rapidity in the CM frame (y^*) by the formula, $y = y^* + y_{cm}$. As can be seen from the Fig. 5.11, the distribution in the LS frame is just shifted by y_{cm} units as compared to the corresponding distribution in the CM frame. Here, $y_{cm} = 8.52$ corresponds to the center-of-mass energy of $\sqrt{s_{NN}} = 5.02\,\text{TeV}$. The shape of the rapidity distribution is observed to be invariant.

Example 5.7 The rapidity formulation simply and naturally incorporates the Lorentz transformation properties of velocity

Solution 5.7 We need to find the particle velocity, v, given frame velocity, $v_0 = 0.8\,c$ and the velocity of the particle in the moving frame is, $v' = 0.6\,c$.

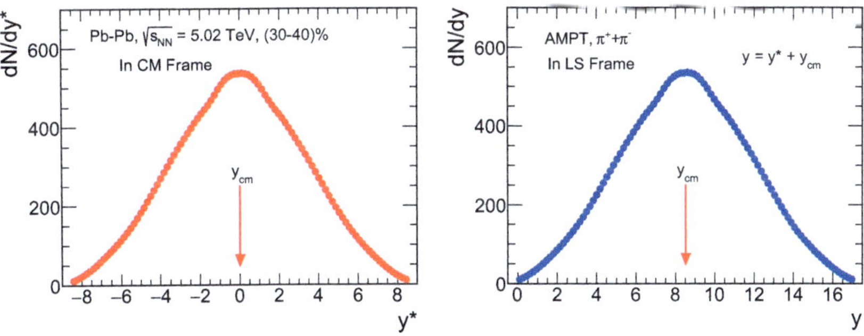

Fig. 5.11 Rapidity distribution of charged pions in the CM and LS frames for Pb-Pb collisions at $\sqrt{s_{NN}} = 5.02$ TeV for (30–30)% centrality class. Invariance of the shape of the rapidity distribution is observed

Now using the inverse Lorentz transformation properties of velocity, one gets

$$v = \frac{v' + v_0}{1 + \left(v_0 v'/c^2\right)}$$

$$= \frac{0.6\,c + 0.8\,c}{1 + 0.6 \times 0.8\,c^2/c^2} = \frac{1.4\,c}{1.48\,c} = 0.945\,c$$

Using the rapidity formulation, one can also obtain the same:

$$y_0 = \frac{1}{2}\ln\left(\frac{1 + \beta_0}{1 - \beta_0}\right)$$

$$= \frac{1}{2}\ln\left(\frac{1 + 0.8}{1 - 0.8}\right) = 1.098$$

Similarly,

$$y' = \frac{1}{2}\ln\left(\frac{1 + 0.6}{1 - 0.6}\right) = 0.693$$

Now

$$y = y' + y_0$$

$$= 1.098 + 0.693 = 1.791$$

Hence, $\beta = \tanh y = \tanh(1.791) = 0.945$.

This explicitly shows that the rapidity formulation naturally incorporates the Lorentz transformation properties of velocity.

Example 5.8 Invariant Cross-section in Terms of Missing Mass and Mandelstam t-variable in an Inclusive Interaction

In lower GeV energies, missing mass analysis is very common. Let us consider a fixed target inclusive interaction: $a + b \rightarrow c + X$. Here, X can be anything in the final state, a is the projectile, and b is the target. The interaction produces c and an unknown particle X. Our aim is to have an expression for the mass, M, of the missing/undetected particle and express the invariant cross-section in terms of Mandelstam t-variable and the missing mass, M.

Solution 5.8 In the laboratory system, one can write the Mandelstam t-variable (squared momentum transfer between target, b and c) as:

$$
\begin{aligned}
t &= (p_b - p_c)^2 \\
&= m_b^2 + m_c^2 - 2m_b E_c
\end{aligned}
\tag{5.124}
$$

The square of the missing mass can be estimated as:

$$
\begin{aligned}
M^2 &= (p_a + p_b - p_c)^2 \\
&= (p_a + p_b)^2 + p_c^2 - 2p_c(p_a + p_b) \\
&= s + m_c^2 - 2E_c(E_a + E_b) + 2\mathbf{p}_c \cdot (\mathbf{p}_a + \mathbf{p}_b) \\
&= s + m_c^2 - 2E_c(E_a + m_b) + 2\mathbf{p}_c \cdot \mathbf{p}_a
\end{aligned}
\tag{5.125}
$$

The last line simplification is because of $\mathbf{p}_b = 0$ (for fixed target experiment- target is at rest). In the CM system, t and M^2 can be written as follows:

$$
t = m_b^2 + m_c^2 - 2E_b^* E_c^* + 2p_{b,z}^* p_{c,z}^*
\tag{5.126}
$$

$$
M^2 = s + m_c^2 - 2\sqrt{s}.E_c^*
\tag{5.127}
$$

Here $p_{c,z}^*$ is the z-component of the momentum of p_c: the longitudinal component, i.e., parallel to the beam direction. Further,

$$
d^3 p = 2\pi \; p_T dp_T \; dp_z^*
\tag{5.128}
$$

One can now change the variables to t and M^2 using appropriate Jacobian,

$$
\begin{aligned}
dt\, dM^2 &= \frac{d\left(t, M^2\right)}{d\left(p_T, p_z^*\right)} \cdot dp_T \, dp_z^* \\[2mm]
&= \begin{vmatrix} \dfrac{dt}{dp_T} & \dfrac{dM^2}{dp_T} \\[2mm] \dfrac{dt}{dp_z^*} & \dfrac{dM^2}{dp_z^*} \end{vmatrix} \cdot dp_T dp_z^*
\end{aligned}
\tag{5.129}
$$

The components of the above Jacobian determinant can be evaluated as follows, which will give the final transformation of the variables. We know in the CM system, the total energy of the particle c is given by:

$$[E_c^*]^2 = p_{T,c}^2 + [p_{z,c}^*]^2 + m_c^2,$$

$$\Rightarrow \frac{dE_c^*}{dp_{T,c}} = \frac{p_{T,c}}{E_c^*} \quad \text{and}$$

$$\frac{dE_c^*}{dp_{z,c}^*} = \frac{p_{z,c}^*}{E_c^*}$$

Using Eqs. (5.126) and (5.127), one obtains:

$$\frac{dt}{dp_{T,c}} = -2E_b^* \cdot \frac{p_{T,c}}{E_c^*} \quad \text{and}$$

$$\frac{dt}{dp_{z,c}^*} = 2p_{b,z}^* - 2E_b^* \cdot \frac{p_{z,c}^*}{E_c^*}$$

$$\frac{dM^2}{dp_{T,c}} = -2\sqrt{s} \cdot \frac{p_{T,c}}{E_c^*} \quad \text{and}$$

$$\frac{dM^2}{dp_{z,c}^*} = -2\sqrt{s} \cdot \frac{p_{z,c}^*}{E_c^*}$$

Hence, the Jacobian becomes:

$$\frac{d\left(t, M^2\right)}{d\left(p_T, p_z^*\right)} = 4p_{b,z}^* \sqrt{s} \cdot \frac{p_{T,c}}{E_c^*} \tag{5.130}$$

Using Eq. (5.129),

$$dt \ dM^2 = \frac{4p_b^* \sqrt{s}}{E_c^*} \cdot p_{T,c} \ dp_{T,c} \ dp_{z,c}^* \tag{5.131}$$

We now need to establish a relationship between dy and dp_z. We know

$$y = \ln\left(\frac{E + p_z}{\sqrt{m^2 + p_T^2}}\right)$$

$$= \ln(E + p_z) - \frac{1}{2}\ln(m^2 + p_T^2) \tag{5.132}$$

Now

$$\frac{dy}{dp_z} = \frac{d}{dp_z}\left[\ln(E + p_z)\right]$$

$$= \frac{1}{(E + p_z)}\left[\frac{dE}{dp_z} + \frac{dp_z}{dp_z}\right] \tag{5.133}$$

We know, $E^2 = p_T^2 + p_z^2 + m^2$. Hence, one obtains $dE/dp_z = p_z/E$. Substituting this in the above equation, we get,

$$dy = \frac{dp_z}{E} \tag{5.134}$$

In our case, $d^3 p = 2\pi \; p_{T,c} dp_{T,c} \, dp_{z,c}^*$. With the use of Eq. (5.131), we get

$$d^3 p = 2\pi \; \frac{E_c^*}{4p_{b,z}^* \sqrt{s}} \, dt \, dM^2$$

$$\Rightarrow \frac{d^3 p}{E_c^*} = \frac{\pi}{2p_{b,z}^* \sqrt{s}} dt \, dM^2$$

$$\Rightarrow E_c^* \frac{d\sigma}{d^3 p} = \frac{2p_{b,z}^* \sqrt{s}}{\pi} \cdot \frac{d\sigma}{dt dM^2} \tag{5.135}$$

$$\simeq \frac{s}{\pi} \cdot \frac{d\sigma}{dt dM^2} \tag{5.136}$$

To obtain the last equation, we have used the relation: $\sqrt{s} \simeq 2p_{b,z}^*$, which is valid for large s.

5.2.3   The Pseudorapidity Variable

Let us assume that a particle is emitted at an angle θ relative to the beam axis. Then its rapidity can be written as:

$$y = \frac{1}{2} \ln\left(\frac{E + p_L}{E - p_L}\right)$$

$$= \frac{1}{2} \ln\left(\frac{E + p_z}{E - p_z}\right)$$

$$= \frac{1}{2} \ln\left[\frac{\sqrt{m_0^2 + p^2} + p\cos\theta}{\sqrt{m_0^2 + p^2} - p\cos\theta}\right]. \tag{5.137}$$

At very high energy, $p \gg m_0$ and hence

$$y = \frac{1}{2} \ln \left[\frac{p + p \cos \theta}{p - p \cos \theta} \right]$$

$$= - \ln(\tan \theta / 2) \equiv \eta \tag{5.138}$$

η is called the pseudorapidity. Hence, at very high energy,

$$y \approx \eta = - \ln(\tan \theta / 2). \tag{5.139}$$

In terms of the momentum, η can be re-written as

$$\eta = \frac{1}{2} \ln \left[\frac{|\mathbf{p}| + p_z}{|\mathbf{p}| - p_z} \right]. \tag{5.140}$$

θ is the only quantity to be measured for the determination of pseudorapidity, independent of any particle identification mechanism. Pseudorapidity is defined for any value of mass, momentum, and energy of the collision. This could also be measured with or without momentum information which needs a magnetic field. A plot of pseudorapidity as a function of the polar angle, θ is shown in Fig. 5.12. Table 5.3 shows the values of η corresponding to the polar angle of emission of a particle. One speaks of the "forward" direction in a collider experiment, which refers to regions of the detector that are close to the beam axis, at high $|\eta|$. The difference in the rapidity of two particles is independent of Lorentz boosts along the beam axis. Pseudorapidity is odd about $\theta = 90°$. In other words, $\eta(\theta) = -\eta(180 - \theta)$ (Fig. 5.13).

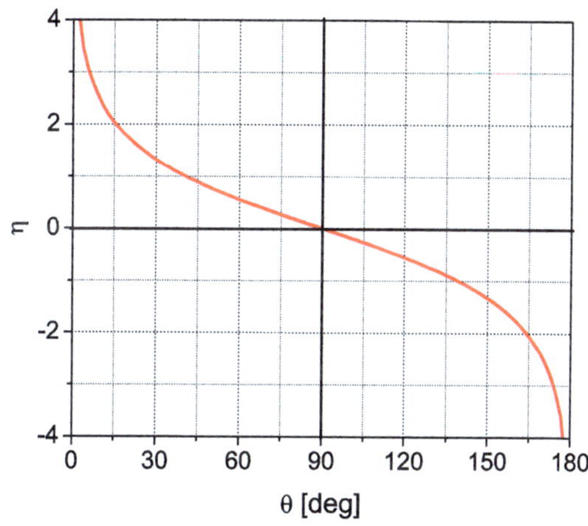

Fig. 5.12 A plot of pseudorapidity variable, η as a function of the polar angle, θ

Table 5.3 Table of pseudorapidity, η vs the polar angle, θ

θ	0	5	10	20	30	45	60	80	90	100	110	...	175	180
η	∞	3.13	2.44	1.74	1.32	0.88	0.55	0.175	0	-0.175	-0.55	...	-3.13	$-\infty$

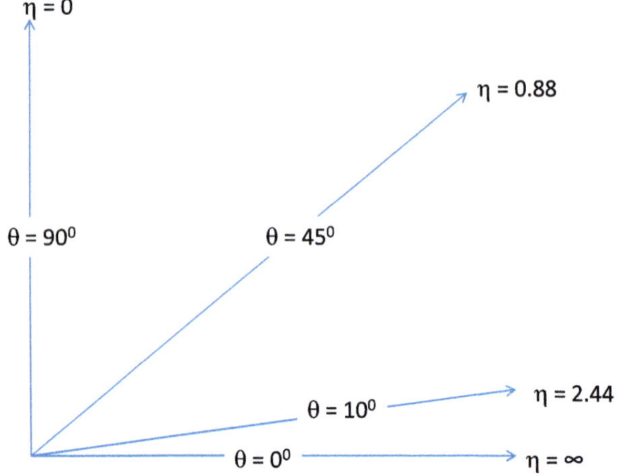

Fig. 5.13 As the angle increases from zero, pseudorapidity decreases from infinity

5.2.3.1 Some Properties of Pseudorapidity

1. The polar angle, θ, is the only quantity to be measured for the estimation of pseudorapidity.
2. No particle identification (PID) is necessary, which makes life easier.
3. Pseudorapidity could be measured with and without the momentum information, which needs a magnetic field.

5.2.3.2 Change of Variables from (y, p_T) to (η, p_T)

By Eq. (5.140),

$$e^{\eta} = \sqrt{\frac{|\mathbf{p}| + p_z}{|\mathbf{p}| - p_z}} \tag{5.141}$$

$$e^{-\eta} = \sqrt{\frac{|\mathbf{p}| - p_z}{|\mathbf{p}| + p_z}} \tag{5.142}$$

Adding both of the equations, we get

$$|\mathbf{p}| = p_T \cosh \eta \tag{5.143}$$

$\mathbf{p}_T = \sqrt{|\mathbf{p}|^2 - p_z^2}$. By subtracting the above equations, we get

$$p_z = p_T \sinh \eta \tag{5.144}$$

Using these equations in the definition of rapidity, we get

$$y = \frac{1}{2} \ln \left[\frac{\sqrt{p_T^2 \cosh^2 \eta + m_0^2} + p_T \sinh \eta}{\sqrt{p_T^2 \cosh^2 \eta + m_0^2} - p_T \sinh \eta} \right] \tag{5.145}$$

Similarly η could be expressed in terms of y as,

$$\eta = \frac{1}{2} \ln \left[\frac{\sqrt{m_T^2 \cosh^2 y - m_0^2} + m_T \sinh y}{\sqrt{m_T^2 \cosh^2 y - m_0^2} - m_T \sinh y} \right] \tag{5.146}$$

The distribution of particles as a function of rapidity is related to the distribution as a function of pseudorapidity by the formula

$$\frac{dN}{d\eta d\mathbf{p}_T} = \sqrt{1 - \frac{m_0^2}{m_T^2 \cosh^2 y}} \frac{dN}{dy d\mathbf{p}_T}. \tag{5.147}$$

In the region $y \gg 0$, the pseudorapidity distribution ($dN/d\eta$) and the rapidity distribution (dN/dy) which are essentially the $\mathbf{p}_T$-integrated values of $\frac{dN}{d\eta d\mathbf{p}_T}$ and $\frac{dN}{dy d\mathbf{p}_T}$ respectively, are approximately the same. In the region $y \approx 0$, there is a small *"depression"* in $dN/d\eta$ distribution compared to dN/dy distribution due to the above transformation. At very high energies where dN/dy has a mid-rapidity plateau, this transformation gives a small dip in $dN/d\eta$ around $\eta \approx 0$ (see Fig. 5.14). However, for a massless particle like a photon, the dip in $dN/d\eta$ is not expected (which is clear from the above equation). Figure 5.15 shows the $dN_{ch}/d\eta$ distributions for π, K, p and photons for Pb+Pb collisions at $\sqrt{s_{NN}} = 5.02$ TeV using AMPT simulation. When a mass-dependent dip is observed for the charged particles, the same is not seen for photons.

Independent of the frame of reference where η is measured, the difference in the maximum magnitude of $dN/d\eta$ appears due to the above transformation. In the CMS, the maximum of the distribution is located at $y \approx \eta \approx 0$ and the η-distribution is suppressed by a factor $\sqrt{1 - m^2/<m_T^2>}$ with reference to the rapidity distribution. In the laboratory frame, however, the maximum is located around half of the beam rapidity $\eta \approx y_b/2$, and the suppression factor is $\sqrt{1 - m^2/<m_T^2>} \cosh^2(y_b/2)$, which is about unity. Given the fact that the shape of the rapidity distribution is independent of frame of reference, the peak

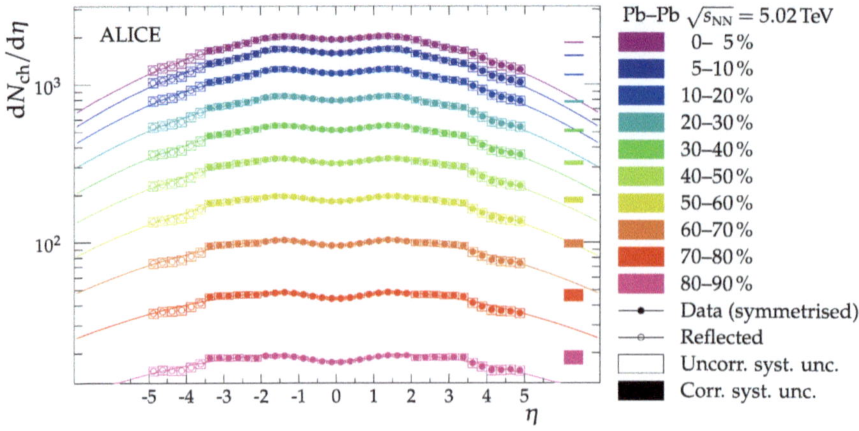

Fig. 5.14 ALICE result of mid-rapidity $dN_{ch}/d\eta$ for Pb+Pb collisions at $\sqrt{s_{NN}} = 5.02$ TeV. Reprinted under CC-BY-4.0 license from [7]. ©2017, ALICE Collaboration

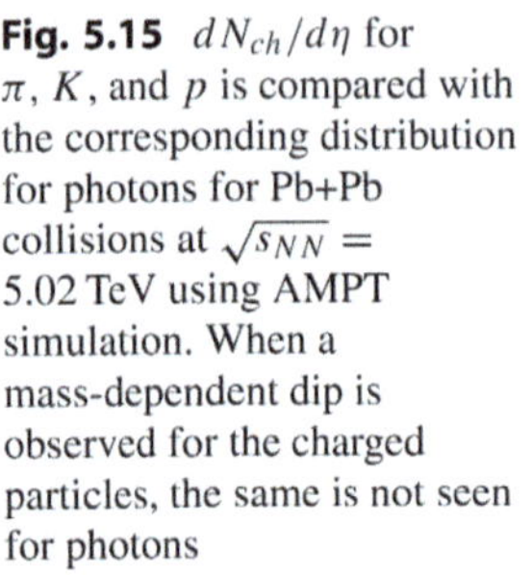

Fig. 5.15 $dN_{ch}/d\eta$ for π, K, and p is compared with the corresponding distribution for photons for Pb+Pb collisions at $\sqrt{s_{NN}} = 5.02$ TeV using AMPT simulation. When a mass-dependent dip is observed for the charged particles, the same is not seen for photons

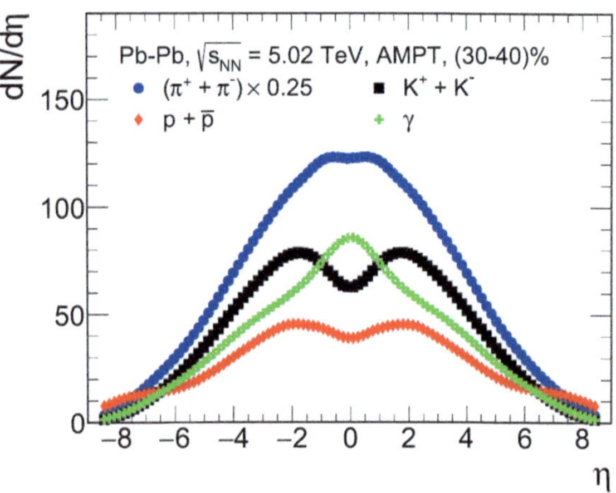

value of the pseudorapidity distribution in the CMS frame is lower than its value in LS. This suppression factor at SPS energies is $\sim 0.8 - 0.9$.

It is worth noting here again that the conversion of rapidity to pseudorapidity phase space is associated with a Jacobian $J(y, \eta)$, which is given by the right-hand side multiplier of Eq. (5.147). This depends on the momentum distribution of the produced particles. In the limit of the rest mass of the particles being much smaller than their momenta, $J(y, \eta) = 1$. The value of the Jacobian is smaller at LHC energies, compared to that at RHIC energies, as the average transverse momentum of particles increases with beam energy. As measured by the PHENIX experiment, for central Au+Au collisions at $\sqrt{s_{NN}} = 200$ GeV, $J(y, \eta) = 1.25$. Whereas, the corresponding measurement of $J(y, \eta) = 1.09$ for Pb+Pb central collisions at $\sqrt{s_{NN}} = 2.76$ TeV by the CMS experiment at LHC. Rewriting Eq. (5.147) after taking an integration over p_T, one obtains:

$$\frac{dN}{d\eta} = v(y)\frac{dN}{dy} \tag{5.148}$$

where $\mathbf{v}$ is the velocity of the particle. For a hadron of mass m_0 and momentum $\mathbf{p}$, which emerges at an angle 90^0 with respect to the beam direction, $y = \eta = 0$, the above relationship becomes,

$$\frac{dN}{d\eta}\Big|_{\eta=0} = v\frac{dN}{dy}\Big|_{y=0} \tag{5.149}$$

As most of the particles are pions in the final state, with an average momentum of three times the pion mass, we have $v = 0.95$. At mid-rapidity, the rapidity-pseudorapidity conversion hence involves almost a constant factor. Hence, the shape is not affected to a greater extent. However, when one considers the whole rapidity range, where the particle velocity in fact becomes a function of the rapidity, the shape of the pseudorapidity distribution (which is characterized by the width) is different from the rapidity distribution.

Usually, the rapidity spectra are parametrized by the sum of two Gaussian distributions positioned symmetrically with respect to mid-rapidity [8, 9],

$$\frac{dN}{dy} = \frac{\langle N \rangle}{2\sqrt{2\pi\sigma^2}}\left\{\exp\left[-\frac{1}{2}\left(\frac{y - y_0}{\sigma}\right)^2\right] + \exp\left[-\frac{1}{2}\left(\frac{y + y_0}{\sigma}\right)^2\right]\right\} \tag{5.150}$$

where $\langle N \rangle$, σ and y_0 are fit parameters. σ is the width of the rapidity distribution. Landau's energy-dependent Gaussian rapidity distribution is given by [10, 11]

$$\frac{1}{\sigma_{in}}\frac{d\sigma}{dy} = \frac{dN}{dy}$$

$$= \frac{N}{(2\pi L)^{1/2}}\exp.\left(-y^2/2L\right) \tag{5.151}$$

with parameters

$$L = \frac{1}{2}\ln\left(s/4m_p^2\right)$$

$$= \ln \gamma = \ln\left(\sqrt{s_{NN}}/2m_p\right) \tag{5.152}$$

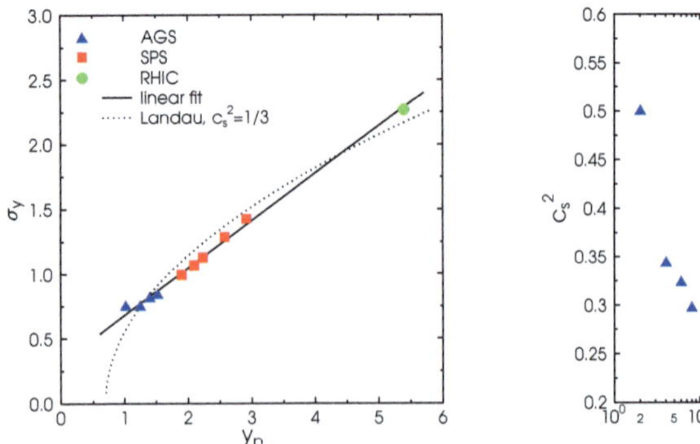
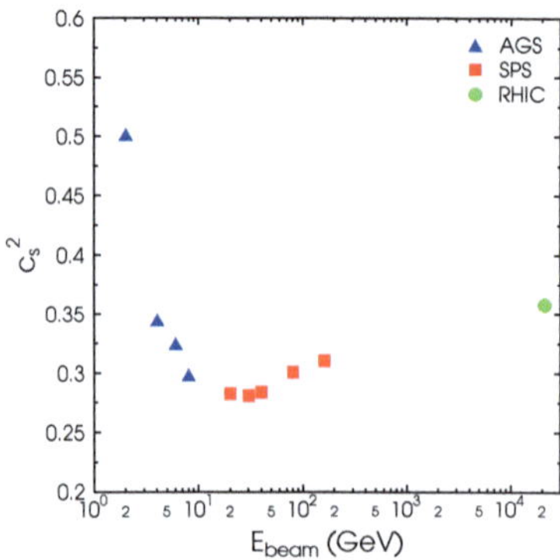

Fig. 5.16 Left: The width of rapidity distributions of π^- in central Pb+Pb (Au+Au) collisions as a function the beam rapidity. The dotted line indicates Landau model predictions with $c_s^2 = 1/3$, while the full line shows a linear fit to data points. Right: Speed of sound as a function of beam energy showing a softest point at $E_{beam} = 30$ AGeV. Reprinted with permission under CC-BY-NC-SA-4.0 license from [12]. ©2006, The Author(s)

where $s \equiv$ squared total center-of-mass energy. Comparing Eqs. (5.150) and (5.151), we get

$$\sigma_y = \sqrt{\ln(\sqrt{s_{NN}}/2m_p)} \qquad (5.153)$$

σ_y is the width of the rapidity distribution, with m_p being the mass of proton (Fig. 5.16). Interestingly, the width of the rapidity distribution is related to the longitudinal flow and velocity of sound in the medium, and hence can probe the equation of state of the produced matter. However, this is affected by the final state rescattering, which could be understood through a p_T dependent study of σ_y to disentangle the initial hard scattering from the final state rescattering. With the assumption that the velocity of sound c_s is independent of temperature, the rapidity density in the framework of the Landau hydrodynamic model is given by [13–16]

$$\frac{dN}{dy} = K \frac{s_{NN}^{1/4}}{\sqrt{2\pi\,\sigma_y^2}} \exp.\left(-\frac{y^2}{2\sigma_y^2}\right) \qquad (5.154)$$

where

$$\sigma_y^2 = \frac{8}{3}\frac{c_s^2}{1-c_s^4}\ln\left(\sqrt{s_{NN}}/2m_p\right)$$

$$\Rightarrow \sigma_y^2 = \frac{8}{3}\frac{c_s^2}{1-c_s^4}\ln\gamma \qquad (5.155)$$

and $K \equiv$ normalization factor. Inverting the above equation for σ_y^2, one can get [12]

$$c_s^2 = \frac{-4}{3} \frac{\ln\left(\sqrt{s_{NN}}/2m_p\right)}{\sigma_y^2} + \sqrt{\left[\frac{4}{3} \frac{\ln\left(\sqrt{s_{NN}}/2m_p\right)}{\sigma_y^2}\right]^2 + 1} \qquad (5.156)$$

For an ideal gas (Landau model prediction), the velocity of sound is $c_s^2 = 1/3$. However, for a gas of hadrons, $c_s^2 = 1/5$. This implies, the expansion of the system is slower compared to an ideal gas scenario. The equation of state is given by

$$\boxed{\frac{\partial P}{\partial \epsilon} = c_s^2} \qquad (5.157)$$

where, P is the pressure and ϵ is the energy of the system under consideration. It should be noted here that when the expansion of the matter proceeds as longitudinal and superimposed transverse expansions, a rarefaction wave moves radially inwards with the velocity of sound. The velocity of sound in the medium formed in heavy-ion collisions, when studied as a function of collision (beam) energy, shows the softest point occurring around $E_{beam} = 30\,\text{AGeV}$. This could be a signature of the deconfinement transition [12].

5.2.4 The Invariant Yield and Transverse Momentum Spectra

The following discussions would be helpful in understanding the invariant yield used in experimental measurement, and hence the transverse momentum spectra, and their connection with the cross-section and luminosity of the accelerator.

1. Cross-sections are the probabilities of nuclear reactions expressed by effective areas. This has a very good discussion, both in Nuclear Physics and Classical Mechanics. We only take up some of the useful aspects of cross-section and differential cross-sections, leaving aside the detailed mathematical derivations, which are not important in this context.
2. The cross-section, σ, is measured in the units of barn or millibarn (1 barn = 10^{-24} cm^2 = 100 fermi^2, 1 mb =0.1 fermi^2). This describes the total yield of a reaction regardless of the energies of the emitted particles or of their spatial distributions.
3. Differential Cross-Section: For example, $\frac{d\sigma}{dE}$ (mb/GeV) or $\frac{d\sigma}{d\theta}$ (mb/radian) are used to study the energy and spatial distributions of the emitted particles, respectively
4. Double Differential Cross-Section: For example, $\frac{d^2\sigma}{dEd\theta}$ - at a given E and θ. d^2N number of particles are emitted into an angular region between θ and $\theta + d\theta$, whose energies lay between E and $E + dE$.

5. Subdivisions of nuclear cross-sections into smaller parts, or in other words, making it a higher-order differential cross-section, would require what information that needs to be derived from the study/measurement. Here, one may consider taking independent variables like v, E, p_T, θ, y, η, etc. to study the reaction cross-section.

6. σ is a Lorentz Invariant quantity. However, differential cross-sections may or may not be invariant. For instance, $\frac{d\sigma}{dE}$ and $\frac{d\sigma}{d\Omega}$.

7. As the Lorentz transformation of common differential cross-sections is more often cumbersome and difficult, one thus needs the use of an invariant cross-section.

8. We know that the energy, E and the momentum, p taken independently are not Lorentz Invariants. However, a suitable combination of E and p, i.e., the energy-momentum four-vector, is Lorentz Invariant. The length of $E - p$ four-vector is Lorentz Invariant, as it is equal to the rest mass of a particle.

$$\sqrt{E^2 - \mathbf{p}^2} = m_0^2. \tag{5.158}$$

9. Another well-known example could be the proper time, $\Delta\tau$,

$$\Delta\tau^2 = \Delta z^2 - (c\Delta t^2). \tag{5.159}$$

is Lorentz Invariant. Here, the halflife, Δt, and the distance travelled, Δz, are not individually Lorentz Invariants.

10. The four-dimensional length element, ds^2, is also Lorentz Invariant, whereas a distance in time and Euclidean space are not individually Lorentz Invariant.

$$ds^2 = c^2 dt^2 - dx^2 - dy^2 - dz^2. \tag{5.160}$$

11.

$$\frac{d^3\sigma}{dp_x dp_y dp_z} = \frac{d^3\sigma}{dp^3}, \tag{5.161}$$

where dp^3 is the elementary volume in momentum space. As the changes in E and $\frac{d^3\sigma}{dp^3}$ cancel out under Lorentz transformation, $E\frac{d^3\sigma}{dp^3}(\equiv \sigma_{inv})$ is Lorentz Invariant.

The rapidity variable has the useful property that it transforms linearly under a Lorentz transformation, so that the invariant differential single particle inclusive cross-section becomes:

$$\boxed{\frac{E d^3\sigma}{dp^3} = \frac{E d^3\sigma}{p_T dp_T dp_L d\phi} = \frac{d^3\sigma}{p_T dp_T dy d\phi}}, \tag{5.162}$$

where

$$dy = \frac{dp_L}{E} \tag{5.163}$$

(This is because: $p_z = m_T \sinh y$, $dp_z = m_T \cosh y \, dy = p_0 \, dy$.) First, we proceed to show $\frac{d^3p}{E}$ is Lorentz invariant. The differential of Lorentz boost in the longitudinal direction is given by

$$dp_z^* = \gamma(dp_z - \beta dE). \tag{5.164}$$

Taking the derivative of the equation $E^2 = p^2 + m^2$, we get

$$EdE = p_z dp_z. \tag{5.165}$$

Using Eqs. (5.164) and (5.165) we get

$$\begin{aligned}
dp_z^* &= \gamma(dp_z - \beta \frac{p_z dp_z}{E}) \\
&= \frac{dp_z}{E}[\gamma(E - \beta p_z)] \\
&= \frac{dp_z}{E} E^*.
\end{aligned} \tag{5.166}$$

As $\mathbf{p}_T$ is Lorentz invariant, multiplying $\mathbf{p}_T$ on both the sides and re-arranging gives

$$\frac{d^3p^*}{E^*} = \frac{d^2p_\perp dp_z^*}{E^*} = \frac{d^2\mathbf{p}_T \, dp_z}{E} = \frac{d^3p}{E}. \tag{5.167}$$

In terms of experimentally measurable quantities, $\frac{d^3p}{E}$ could be expressed as

$$\begin{aligned}
\frac{d^3p}{E} &= dp_x dp_y \left(\frac{dp_z}{E}\right) \\
&= (dp_x dp_y)\, dy \\
&= d\mathbf{p}_T \, dy \\
&= p_T dp_T d\phi dy \tag{5.168} \\
&= m_T dm_T d\phi dy. \tag{5.169}
\end{aligned}$$

For a better visualization, the transverse area element is shown in the Fig. 5.17. This can also be mathematically understood from the Jacobian transformation of variables, as shown below.

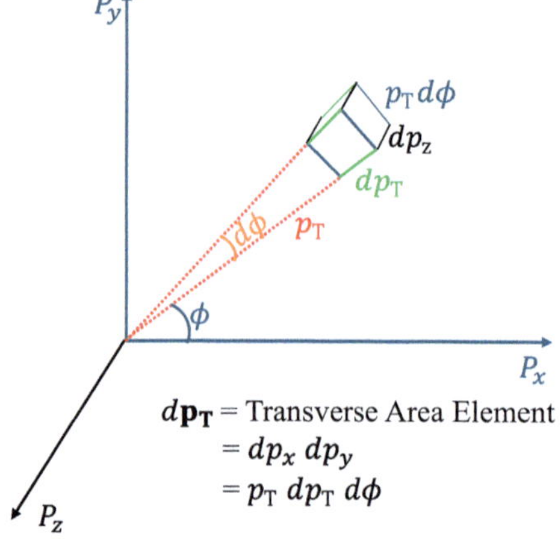

Fig. 5.17 Taking the Lorenz boost along the assumed longitudinal direction, which is the Z-direction, the transverse plane is XY and hence the transverse area element in momentum space, $d\mathbf{p}_T$ can be written as $p_T dp_T d\phi$

We have the transverse momentum,

$$p_T = \sqrt{p_x^2 + p_y^2}$$

$$p_x = p_T \cos\phi \tag{5.170}$$

$$p_y = p_T \sin\phi$$

where, ϕ is the angle p_T makes with the p_x-axis. A coordinate transformation from (p_x, p_y) to (p_T, ϕ) gives:

$$dp_x dp_y = |J|\, p_T d\phi, \tag{5.171}$$

where the Jacobian J is given by:

$$J = \begin{vmatrix} \frac{\partial p_x}{\partial p_T} & \frac{\partial p_x}{\partial \phi} \\ \frac{\partial p_y}{\partial p_T} & \frac{\partial p_y}{\partial \phi} \end{vmatrix} = \begin{vmatrix} \cos\phi & -p_T \sin\phi \\ \sin\phi & p_T \cos\phi \end{vmatrix} = p_T \cos^2\phi + p_T \sin^2\phi \tag{5.172}$$

$$= p_T$$

$$\Rightarrow dp_x dp_y = p_T dp_T d\phi.$$

Hence, the area element in the transverse plane is given by:

$$d\mathbf{p}_T = p_T dp_T d\phi. \tag{5.173}$$

In high-energy physics involving multiparticle production processes, one, however, measures the invariant yield, $\frac{Ed^3N}{dp^3}$. In terms of experimentally measurable quantities, this could be expressed as

$$
\begin{aligned}
\frac{Ed^3N}{dp^3} &= \frac{1}{m_T}\frac{d^3N}{dm_T d\phi dy} \\
&= \frac{1}{2\pi\, m_T}\frac{d^2N}{dm_T dy} \\
&= \frac{1}{2\pi\, p_T}\frac{d^2N}{dp_T dy}.
\end{aligned}
\tag{5.174}
$$

To have a statistically significant number in the differential yield, one combines many events of similar nature, e.g., say central events or peripheral events or events of a particular centrality class. This requires at the end, an event normalization, which makes the above equation,

$$
\frac{Ed^3N}{dp^3} = \frac{1}{N_{event}}\frac{1}{2\pi\, p_T}\frac{d^2N}{dp_T dy}
\tag{5.175}
$$

To measure the invariant yields of identified particles, Eq. (5.175) is used by experiments. Figure 5.18 (upper) shows the experimental invariant yield of charged pions, kaons, and protons(antiprotons) for Pb-Pb collisions at $\sqrt{s_{NN}} = 2.76\,\text{TeV}$ and Au-Au collisions at $\sqrt{s_{NN}} = 200\,\text{GeV}$. In the lower figure, the invariant yields of various identified particles are shown for minimum-bias pp collisions at $\sqrt{s} = 13\,\text{TeV}$. One can observe here that the factor $2\pi p_T$ is not there in the denominator of the invariant yield formula, as is seen in the upper figure. This is sometimes taken care of by multiplying it with the obtained numbers, dN for a given $dp_T dy$-bin. The figures here additionally demonstrate a mass hierarchy in the particle production, meaning a thermalized production of particles following a Boltzmann distribution with a preference for low-mass particles to populate the mass-spectrum abundantly.

As can be easily observed from the above discussions on the invariant yield, the use of p_T or m_T doesn't make any fundamental difference to the spectra, except for the m_T-distribution of an individual spectrum on the m_T-axis is shifted depending on the rest mass of the particle. This is because $m_T^2 = p_T^2 + m_0^2$. For $p_T = 0$, a m_T-spectrum starts from m_0, the rest mass of a particle. This shift can be seen in Fig. 5.19, where individual spectrum starts from different positions depending on the rest masses, instead of starting from zero or the lowest measurable p_T of the particle.

The invariant cross-section and the invariant yield are not necessarily the same, as it is used more frequently in an erroneous way. One ends up with a dimension crisis while trying to equalize the both. The connection between the two quantities is

Fig. 5.18 Invariant yield distribution for Pb-Pb collisions at $\sqrt{s_{NN}} = 2.76\,\text{TeV}$ and Au-Au collisions at $\sqrt{s_{NN}} = 200\,\text{GeV}$ (Top). Reprinted under CC-BY-3.0 license from [18]. ©2012, ALICE Collaboration. The bottom figure shows invariant yields of various identified particles for minimum-bias pp collisions at $\sqrt{s} = 13\,\text{TeV}$. Reprinted under CC-BY-4.0 license from [19]. ©2021, ALICE Collaboration

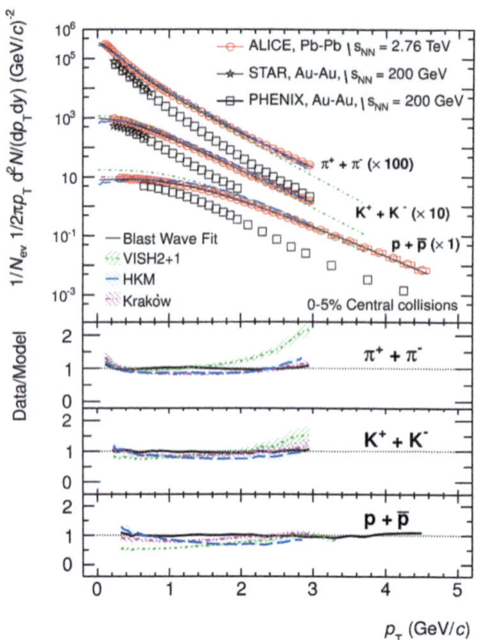

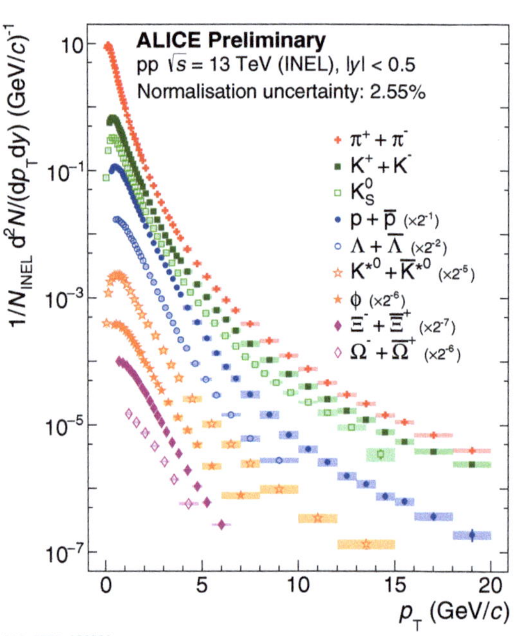

Fig. 5.19 Invariant yield distribution (using m_T) for various identified particles for minimum-bias pp collisions at $\sqrt{s} = 13\,\mathrm{TeV}$. Reprinted under CC-BY-4.0 license from [19]. ©2021, ALICE Collaboration

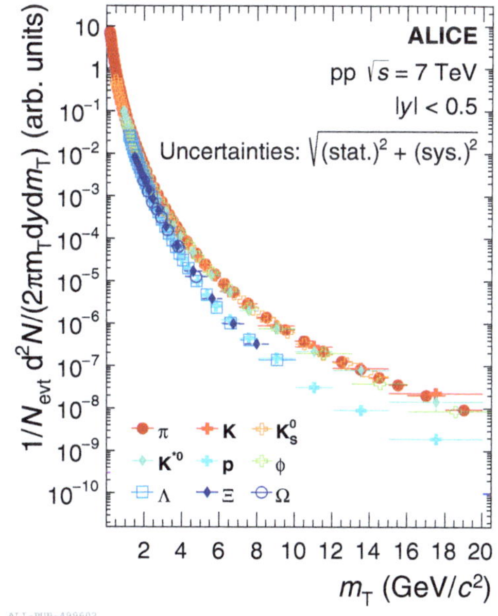

as follows. When we calculate the differential cross-section from differential yield, we need to divide the integrated luminosity by the differential yield,

$$\frac{d^2\sigma}{2\pi p_T dp_T dy} = \frac{1}{\mathcal{L}_{int}} \frac{d^2 N}{2\pi p_T dp_T dy}. \tag{5.176}$$

N is the number of particles and $\mathcal{L}_{int}$ is the integrated luminosity, given by $\mathcal{L}_{int} = \int \mathcal{L}_{inst}\, dt$, where $\mathcal{L}_{inst}$ is the instantaneous luminosity. The unit of $\mathcal{L}_{int}$ is mb^{-1}.

Example 5.9 Use Jacobian method to show that $\frac{d^3 p}{E}$ is an invariant quantity

Solution 5.9 The quantity $\frac{d^3 p}{E}$ is also called as "Phase Space Volume Element". Consider the four momenta in a Lorentz frame moving with velocity β with respect to the laboratory frame as: $(E^*, p_x^*, p_y^*, p_z^*)$. If the boost is along the longitudinal, z-axis, then one can proceed to write the relationships between the two frames as:

$$E = \gamma(E^* + \beta p_z^*) \tag{5.177}$$

$$p_z = \gamma(p_z^* + \beta E^*) \tag{5.178}$$

$$p_x = p_x^* \tag{5.179}$$

$$p_y = p_y^* \tag{5.180}$$

Using the Jacobian transformation of variables:

$$dp_x dp_y dp_z = \frac{d(p_x, p_y, p_z)}{d(p_x^*, p_y^*, p_z^*)} dp_x^* dp_y^* dp_z^* \tag{5.181}$$

Due to the Lorentz transformation, only E and p_z were observed to change. The Jacobian thus becomes

$$\frac{d(p_x, p_y, p_z)}{d(p_x^*, p_y^*, p_z^*)} = \frac{dp_z}{dp_z*} \tag{5.182}$$

$$= \gamma \left(1 + \beta \frac{dE^*}{dp_z^*}\right)$$

$$= \gamma \left(1 + \beta \frac{p_z^*}{E^*}\right)$$

$$= \frac{\gamma(E^* + \beta p_z^*)}{E^*}$$

$$= \frac{E}{E^*} \tag{5.183}$$

Using this in Eq. (5.181), one obtains

$$d^3 p = \frac{E}{E^*} d^3 p^*$$

$$\Rightarrow \frac{d^3 p}{E} = \frac{d^3 p^*}{E^*} \tag{5.184}$$

This proves the invariance of the phase space volume element under Lorentz transformation.

5.2.5   Luminosity

Luminosity is defined as a quantity that measures the ability of a particle accelerator to produce the required number of interactions (Table 5.4). This is an accelerator-specific parameter. The number of useful interactions (events) becomes important, especially when rare events with smaller production cross-sections (σ_p) are studied. The relationship between the rate of the interaction and cross-section is:

$$\frac{dR}{dt} = \mathcal{L}.\sigma_p \tag{5.185}$$

The unit of $\mathcal{L}$ is $cm^{-2}\ s^{-1}$.

Table 5.4 Luminosity of different Particle Accelerators. The asterisk (*) marked accelerators are the upcoming ones. Figure 5.20 shows the peak luminosity as a function of year, depicting the time evolution of accelerator technology [20,21]. The table is made by integrating data from different sources

Accelerator	Location	Beam energy ($E_A + E_B$) (GeV)	$\mathcal{L}\,(cm^{-2}s^{-1})$
LHC (Pb+Pb)	CERN	2510+2510	$\sim 7 \times 10^{26}$
LHC (p+Pb)	CERN	$\sqrt{s_{NN}} = 8160$	3×10^{29}
RHIC (Au+Au)	BNL	100 + 100	5×10^{27}
RHIC (d+Au)	BNL	100.7 + 100	4×10^{29}
FAIR* (A+A)	GSI	$\sqrt{s_{NN}} = 2.7 - 9.3$	
NICA* (A+A)	JINR	$\sqrt{s_{NN}} = 3 - 9$	$\sim 10^{27}$
LHC ($p + p$)	CERN	6500 + 6500	$\sim 6 \times 10^{33}$
RHIC (polarized $p + p$)	BNL	254.9+254.9	7×10^{31}
ISR ($p + p$)	CERN	31+31	6×10^{31}
TEVATRON ($p + \bar{p}$)	FermiLab	1000+1000	50×10^{30}
SPS ($p + \bar{p}$)	CERN	315+315	6×10^{30}
ISR ($\bar{p} + p$)	CERN	31+31	$\sim 10^{27}$
KEKB (e^+e^-)	JAPAN	8+3.5	$\sim 10^{34}$
PEP (e^+e^-)	SLAC	9+3	3×10^{33}
LEP (e^+e^-)	CERN	105 + 105	$\sim 10^{32}$
PETRA (e^+e^-)	DESY	23+23	2×10^{31}
HERA ($e^+ p$)	DESY	30+920	40×10^{30}

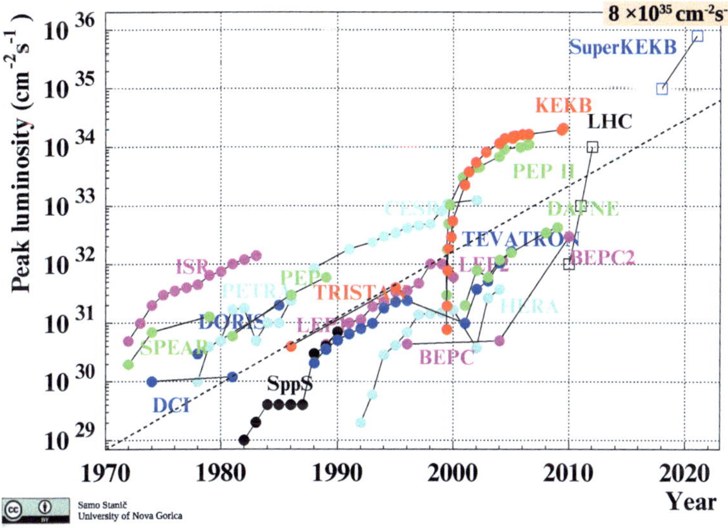

Fig. 5.20 The peak luminosity as a function of year, showing the accelerator technology evolution with time. Reprinted under CC-BY-4.0 license. ©2025, Samo Stanič

5.2.5.1 Luminosity in Fixed Target Experiments

Here, one needs to consider the properties of the incoming beam and the stationary target. For simplicity, we assume a homogeneous thin target characterized by density, ρ (in gm/cm^3) and thickness, l (in cm). The incoming beam is characterized by the beam flux, Φ, defined by the number of particles per unit cross-sectional area per unit time.

- $a \equiv$ area hit by the incoming beam particles.
- $\sigma_p \equiv$ cross-section: each proton representing an effective target area σ_p.
- $n \equiv$ number of target protons per cm^3: $n = \frac{N_{\mathrm{AV}}}{A} . \rho$ (in cm^{-3})

  Here, N_{AV} is Avogadro's number, A is the atomic weight of the target material in gm. Hence:
- Number of nuclei hit by the beam $= nla$.
- probability of interaction $= \frac{\text{beam hit area}}{\text{beam dimension}} = \frac{nla\sigma_p}{a} = nl\sigma_p$.

  This is equivalent to the ratio of (number of events per second) to the number of beam particles per second. This equals the ratio of event rate $(\frac{dR}{dt})$ to N_{beam} per second.
- Event rate thus becomes: $\frac{dR}{dt} = [N_{\mathrm{beam}}. \, n. \, l]. \, \sigma_p = \mathcal{L}_{FT}.\sigma_p$

In terms of beam flux, the luminosity and the interaction rate for a fixed target experiment are respectively given by

$$\mathcal{L}_{FT} = \Phi\rho l. \tag{5.186}$$

The interaction rate:

$$\frac{dR}{dt} = \Phi\rho l.\sigma_p = \mathcal{L}_{FT}.\sigma_p. \tag{5.187}$$

Luminosity, $\mathcal{L}_{FT}$ is measured in $cm^{-2}.sec^{-1}$. It could be easily realized that the luminosity in a fixed target can be increased by using a target material of higher density. To have an analogy, in low-energy nuclear physics or condensed matter physics, people talk about the beam current as one of the characteristics of the incoming beam. The beam flux is the beam current per unit area.

Example 5.10 Luminosity Estimation for a Fixed Target Experiment

In a fixed target experiment, a liquid hydrogen target is used, which is characterized by a target density, $\rho = 0.072 \, gm.cm^{-3}$ with a target thickness of 15 cm. The beam flux is given by $N_{\mathrm{beam}} = 10^{11}$ protons/sec. Estimate the luminosity for this setup.

Solution 5.10 Here the number of target protons is given by: $n = 6 \times 10^{23} \times 0.072 \simeq 0.432 \, cm^{-3} \times 10^{23}$. Hence, the luminosity is given by:

$$\mathcal{L}_{FT} = N_{\mathrm{beam}} \, n. \, l = 10^{11} \times 0.432 \times 10^{23} \times 15 = 6.48 \times 10^{34} \, cm^{-2}.sec^{-1}.$$

5.2.5.2 Luminosity in Collider Experiments

Considering a Gaussian beam profile with head-on collisions, the luminosity for a collider experiment could be written as

$$\mathcal{L} = \frac{f n_1 n_2 n_b}{4\pi \sigma^2},$$

(5.188)

where $\sigma^2 = \sigma_x \sigma_y \equiv$ beam size at the interaction point,

$n_1, n_2 \equiv$ number of particles in bunch 1 and bunch 2,
$n_b \equiv$ number of bunches per beam,
$f \equiv$ bunch revolution frequency.

Integrated Luminosity Is Defined as

$$L = \int \mathcal{L}\, dt.$$

(5.189)

This is expressed in cm^{-2}.

Example 5.11 Effect of Accelerator Luminosity on Event Statistics
The Large Hadron Collider (LHC) is a 40 MHz particle collider. We consider proton+proton collisions at a center-of-mass energy, $\sqrt{s} = 13$ TeV, with 1.15×10^{11} protons/bunch and 2808 bunches/beam with a beam size of 16 μm at the interaction point. Calculate the luminosity of the LHC machine for this configuration. If you are interested in a rare process having cross-section, $\sigma_p = 1$ femtobarn ($=10^{-39}$ cm^2), to bound the statistical errors by 2%, how much time does the LHC have to take data?

Solution 5.11 Here, $n_1 = n_2 = 1.15 \times 10^{11}$ protons/bunch,
  $n_b = 2808$ bunches/beam,
  $\sigma \equiv$ beam size at the interaction point $= 16$ $\mu m = 16 \times 10^{-4}$ cm,
  Given LHC is a 40 MHz particle collider $\Rightarrow$ f$= 40 \times 10^6$ sec^{-1}. Now the luminosity is given by

$$\mathcal{L} = \frac{f n_1 n_2 n_b}{4\pi \sigma^2},$$

$$= \frac{40 \times 10^6 \ sec^{-1} \times (1.15 \times 10^{11})^2 \times 2808}{4\pi \times (16 \times 10^{-4} \ cm)^2}$$

$$- 46\ 168 \times 10^{36} \ cm^{-2} \ sec^{-1}$$

Table 5.5 Comparison between fixed target vs collider

Sl. No.	Fixed target	Collider
1.	$\sqrt{s} \propto \sqrt{E_{\text{lab}}}$	$\sqrt{s} \propto E_{\text{cm}}$
2.	Energy available for particle production is less	Energy available for particle production grows faster
3.	Many kinds of beams are possible using secondary targets: p, π, K, μ	Only stable and charged beams particles: $pp(\bar{p})$, e^+e^-, e^-p, heavy-ions
4.	Many different types of targets are possible	Colliding beam particles as mentioned in (3)
5.	Higher luminosity achievable	Difficult to achieve higher luminosity

Given $\sigma_p = 1\ fb = 10^{-39}\ cm^2$,

$$\frac{dR}{dt} = \mathcal{L} \cdot \sigma_p$$

$$= 10^{-39}\ cm^2 \times 46.168 \times 10^{36}\ cm^{-2}\ sec^{-1}$$

$$= 46.168 \times 10^{-3}\ sec^{-1}$$

$$= 166\ hr^{-1}$$

This means there are 166 events occurring per hour at the LHC. To bound the statistical errors by 2% means,

$$\frac{1}{\sqrt{N}} = 0.02$$

$$\Rightarrow N = \frac{1}{0.02^2} = 2500$$

Hence, to bound statistical errors by 2%, LHC needs to run for: $2500/166 \simeq 15$ hours.

Table 5.5 gives a comparison of luminosities between fixed target and collider experiments.

5.2.6   The Feynman Scaling Variable x_F

Richard Feynman introduced the idea of *scaling variables* to describe particle production at very high energies in terms of dimensionless quantities. Consider a reaction of type:

$$beam\ +\ target\ \longrightarrow\ A\ +\ anything,$$

where A is known is called an *"inclusive reaction"*. The cross-section for particle production could be written separately as functions of $\mathbf{p}_T$ and p_L:

$$\sigma = f(\mathbf{p}_T)g(p_L).$$ (5.190)

This factorization is empirical in nature and convenient because each of these factors has simple parametrizations that fit well to experimental data.

Similarly, the differential cross-section could be expressed by

$$\frac{d^3\sigma}{dp^3} = \frac{d^2\sigma}{d\mathbf{p}_T^2}\frac{d\sigma}{dp_L}$$ (5.191)

Define the variable

$$x_F = \frac{p_L^*}{p_L^*(max)}$$

$$\Rightarrow \boxed{x_F = \frac{2p_L^*}{\sqrt{s}}}$$ (5.192)

x_F is called the *Feynman scaling variable*. Feynman hypothesized that, at asymptotically high energies, the inclusive distribution of produced secondaries in hadronic and heavy-ion interactions, becomes independent of the beam energy when expressed in terms of x_F. This is called *Feynman scaling*.

The longitudinal component of the cross-section when measured in CMS of the collision, would scale i.e. would not depend on the collision energy $\sqrt{s}$. This is the fraction of the maximum allowed longitudinal momentum carried by the particle in the CMS. This variable is used in comparing the shapes of particle distributions at different collision energies near the projectile or target rapidity. Instead of $\frac{d\sigma}{dp_L^*}$, $\frac{d\sigma}{dx_F}$ is measured, which wouldn't depend on the energy of the reaction, $\sqrt{s}$. This assumption of Feynman's is valid approximately.

Note
- The bounds of x_F: $-1 \le x_F \le 1$
- $x_F = 1$: particle moving forward with the maximum possible longitudinal momentum.
- $x_F = -1$: particle moving backward with the maximum possible longitudinal momentum (opposite to the beam direction in a fixed-target experiment).
- $x_F = 0$: particle produced at central rapidities in the transverse region.

Let us now derive this Feynman scaling in high-energy physics. The differential cross-section for the inclusive production of a particle is then written as

$$\frac{d^3\sigma}{dx_F d^2\mathbf{p}_T} = F(s, x_F, \mathbf{p}_T)$$ (5.193)

Feynman's assumption that at high energies the function $F(s, x_F, \mathbf{p}_T)$ becomes asymptotically independent of the energy means:

$$\lim_{s \to \infty} F(s, x_F, \mathbf{p}_T) = F(x_F, \mathbf{p}_T) = f(\mathbf{p}_T) \, g(x_F) \qquad (5.194)$$

According to Feynman, the mean number of particles rises logarithmically in the asymptotic limit of large energies. This in fact, applies to any kind of particles:

$$< N > \; \propto \ln W \; \propto \ln \sqrt{s},$$

where $W = \sqrt{s}/2$, is the beam energy of a symmetric collider. However, these conclusions are based on phenomenological arguments about the exchange of quantum numbers between the colliding particles. It was assumed that the number of particles with a given mass and transverse momentum ($\mathbf{p}_T$) in a longitudinal interval, dp_L, depends on the energy:

$$E = E(p_L)$$

As we take the z-direction as the longitudinal direction along the beam axis, we now replace dp_L with dp_z for consistency. With this,

$$\frac{dN}{dp_z} \sim \frac{1}{E} \qquad (5.195)$$

This means the among all the produced particles, the probability of finding a particle of kind i with transverse momentum $\mathbf{p}_T$, rest mass m_0, and longitudinal momentum p_z is of the form:

$$f_i(\mathbf{p}_T, \; x_F = p_z/W) \, \frac{dp_z}{E} \, d^2 p_T \qquad (5.196)$$

with $E = \sqrt{m_0^2 + p_T^2 + p_z^2}$ being the total energy of the particle under discussion. The function $f_i(\mathbf{p}_T, \; x_F)$ denotes the particle distribution. Let's derive Eq. (5.196) to have a hand on experience.

Rewriting Eq. (5.196) in the form of an invariant cross-section, we get:

$$\frac{1}{\sigma} \, E \frac{d^3 \sigma}{dp_z \, d^2 \mathbf{p}_T} = f_i(\mathbf{p}_T, \; x_F). \qquad (5.197)$$

It is experimentally found that f_i factorizes approximately. We choose a normalization, g_i, such that:

$$\int f_i(\mathbf{p}_T, \; x_F) \, d^2 p_T = f_i(x_F) \int g_i(\mathbf{p}_T) \, d^2 p_T$$

$$= f(x_F). \qquad (5.198)$$

Here,

$$\int g_i(\mathbf{p}_T)\, d^2 p_T \;=\; 1 \tag{5.199}$$

Now, integrating Eq. (5.197) and applying Eq. (5.198), we get

$$\int \frac{1}{\sigma}\, E\, \frac{d^3\sigma}{dp_z\, d^2\mathbf{p}_T}\, \frac{d^3 p}{E} \;=\; <N>$$

$$= \int f_i(\mathbf{p}_T,\, x_F)\, \frac{d^3 p}{E}$$

$$= \int f_i(x_F)\, \frac{dp_z}{\sqrt{W^2 x_F^2 + m_T^2}} \tag{5.200}$$

On the left hand side we have used the definition of invariant cross-section with the average particle multiplicity $<N>$, and for m_T an effective average p_T is used with $x_F = p_z/W$. Hence

$$<N> \;=\; \int_{-1}^{+1} f_i(x_F)\, \frac{dx_F}{\sqrt{x_F^2 + \frac{m_T^2}{W^2}}}, \tag{5.201}$$

with $dx_F = \frac{dp_z}{W}$.

The integral is symmetric because $f_i(x_F)$ is symmetric for collisions of identified particles. For asymmetric collision systems, the integration could be performed separately for negative and positive x_F and yields the same result.

$f_i(x_F) \leq B$ is finite and bounded due to energy conservation, where B is the bound. Feynman assumed that for $x_F = 0$ a finite limit is reached.

$$x_F \to 0 \;\Rightarrow\; \frac{p_z}{W} = \frac{p_z}{E_{lab}} \;\to\; 0$$

$$\Rightarrow E_{lab} \to \infty$$

Hence

$$<N> = 2 \int_0^1 f_i(x_F)\, \frac{dx_F}{\sqrt{x_F^2 + \frac{m_T^2}{W^2}}} \;\leq\; 2 \int_0^1 B\, \frac{dx_F}{\sqrt{x_F^2 + \frac{m_T^2}{W^2}}}$$

$$= 2B\, ln\left[\, x_F + \sqrt{x_F^2 + \frac{m_T^2}{W^2}}\,\right]\Bigg|_0^1$$

$$= 2B \ln \left[1 + \sqrt{1 + \frac{m_T^2}{W^2}} \right] - 2B \ln \left(\frac{m_T}{W} \right) \qquad (5.202)$$

In the limit $W \to \infty$, the first term of the above equation could be shown to be constant, and the second term is proportional to $\ln W$. hence, Feynman scaling tells that the average total multiplicity scales as:

$$\boxed{< N > \ \propto \ \ln W \ \propto \ \ln \sqrt{s}} \qquad (5.203)$$

If one considers the maximum reachable rapidity in a collision to increase like $\ln \sqrt{s}$ ($y_{max} \sim \ln \sqrt{s}$) and in addition, particles are evenly distributed in rapidity, it follows that $\frac{dN}{dy}$ is independent of $\sqrt{s}$:

$$\boxed{\frac{dN}{dy} = Constant} \qquad (5.204)$$

Feynman's Assumption $f_i(p_T, x_F)$ which denotes the particle distribution, becomes independent of W at high energies. This assumption is known as *Feynman Scaling* and f_i is called the scaling function or Feynman function. The variable

$$\boxed{x_F = \frac{p_z}{W} = \frac{2p_z}{\sqrt{s}}} \qquad (5.205)$$

is called **Feynman-x variable**.

Feynman-x is the ratio of the longitudinal momentum of the particle to the total energy of the incident particle.

Note
1. $\frac{dN}{dy}$ = Constant.
   $\Rightarrow$ Height of the rapidity distribution around mid-rapidity (the so-called plateau) is independent of collision energy, $\sqrt{s}$.
2. Equivalently, if Feynman scaling holds good, the pseudorapidity density of charged particles at mid-rapidity i.e. $\frac{dN}{d\eta}(\eta = 0)$, is approximately constant.
   Scaling properties of various distributions can also be studied in terms of the scaled rapidity:

$$\boxed{z = \frac{y^*}{y^*_{max}}} \ . \qquad (5.206)$$

The two scaled variables: x_F and z emphasize different kinematic regions: the detailed structure of the central part of the distribution (i.e. large emission angles) can be better seen using x_F, while the far *"wings"* (i.e. small angles) using z.

5.2.7 Understanding a p_T or m_T Spectrum

The p_T spectrum: Plotting $\frac{1}{2\pi p_T}\frac{d^2N}{dy\,dp_T}$ as a function of p_T.

The m_T spectrum: Plotting $\frac{1}{2\pi m_T}\frac{d^2N}{dy\,dm_T}$ as a function of m_T or $(m_T - m_0)$.

The invariant yield when plotted as a function of p_T or m_T or $(m_T - m_0)$ is called p_T or m_T spectrum, respectively. Experimentally at kinetic freeze-out when the elastic collisions between the final state particles almost cease to happen (in other words the particle mean free path becomes higher than the system size i.e., size of the produced fireball) then the p_T or m_T-spectrum is frozen, which carries the kinetic freeze-out properties of the system. Recall here that:

$$\boxed{p_T dp_T = m_T dm_T}.$$

The distribution of particles as a function of p_T is called p_T-distribution. Mathematically,

$$\frac{dN}{d\mathbf{p}_T} = \frac{dN}{2\pi\ |\mathbf{p}_T|d|\mathbf{p}_T|} \tag{5.207}$$

where dN is the number of particles in a particular p_T-bin. People usually plot $\frac{dN}{p_T dp_T}$ as a function of p_T taking out the factor $1/2\pi$, which is a constant. Here p_T is a scalar quantity. The low-p_T part of the p_T-spectrum is well described by an exponential function having thermal origin given by Eq. (5.208). However, a QCD-inspired power-law function (given by Eq. (5.209)) seems to provide a better description of the high p_T ($\gtrsim 3\,\text{GeV/c}$) region. To describe the whole range of the p_T-spectrum, one uses the Levy function given by Eq. (5.210) which has an exponential part to describe low-p_T and a power-law function to describe the high p_T part, which is dominated by hard scatterings (high momentum transfer at early times of the collision).

$$\frac{1}{2\pi p_T}\frac{d^2N}{dy\,dp_T} = A\,e^{\frac{-m_T}{T}}, \tag{5.208}$$

$$\frac{1}{2\pi p_T}\frac{d^2N}{dy\,dp_T} = B\left(1 + \frac{p_T}{p_0}\right)^{-n}, \tag{5.209}$$

$$\frac{1}{2\pi p_T}\frac{d^2N}{dy\,dp_T} = \frac{dN}{dy}\frac{(n-1)(n-2)}{2\pi nC[nC + m_0(n-2)]} \times \left(1 + \frac{\sqrt{p_T^2 + m_0^2} - m_0}{nC}\right)^{-n}, \tag{5.210}$$

where $A, T, B, p_0, n, \frac{dN}{dy}, C$, and m_0 are fit parameters [??] The inverse slope parameter of p_T-spectra is called the effective temperature (T_{eff}), which has a thermal contribution because of the random kinetic motion of the produced particles and a contribution from the collective motion of the particles.

If we look deeper into Eq. (5.208), we expect this for a 2-dimensional classical thermalized fluid at rest. We recall here that the Boltzmann distribution $\propto \exp(-\beta E)$, where $\beta \equiv 1/T$ is the inverse temperature. We use a semi-logarithmic plot (y-axis logarithmic scale) of $N \exp(-\beta m_T)$ vs $(m_T - m_0)$, which happens to be a straight line with slope $-\beta$, from which the effective temperature is extracted. A semi-log plot of $N \exp(-\beta m_T)$ vs p_T is an approximate straight line if $p_T \gg m_0$.

The most important parameter is then the mean transverse momentum ($\langle p_T \rangle$), which carries the information of the effective temperature of the system. Experimentally, $\langle p_T \rangle$ is studied as a function of $\frac{dN_{ch}}{d\eta}$, which is the measure of the entropy density of the system. This is like studying the temperature as a function of entropy to see the signal of a phase transition. The phase transition is of 1st order if a plateau is observed in the spectrum, signalling the existence of latent heat of the system (like liquid-vapour phase transition). This was first proposed by L. Van Hove [23].

The average of any quantity A following a particular probability distribution $f(A)$ can be written as

$$\langle A \rangle = \frac{\int A\, f(A)\, dA}{\int f(A)\, dA}. \tag{5.211}$$

Similarly,

$$\begin{aligned}
\langle p_T \rangle &= \frac{\int_0^\infty p_T \left(\frac{dN}{dp_T}\right) dp_T}{\int_0^\infty \left(\frac{dN}{dp_T}\right) dp_T} \\[2mm]
&= \frac{\int_0^\infty p_T\, dp_T\ p_T \left(\frac{dN}{p_T dp_T}\right)}{\int_0^\infty p_T\, dp_T \left(\frac{dN}{p_T dp_T}\right)} \\[2mm]
&= \frac{\int_0^\infty p_T\, dp_T\ p_T\, f(p_T)}{\int_0^\infty p_T\, dp_T\, f(p_T)}.
\end{aligned} \tag{5.212}$$

The p_T-distribution function is given by

$$f(p_T) = \frac{dN}{d\mathbf{p}_T} = \frac{dN}{p_T dp_T}. \tag{5.213}$$

Example 5.12 Estimation of $< m_T >$

Solution 5.12 Experimental data on p_T-spectra are sometimes fitted to the exponential Boltzmann-type function given by

$$f(p_T) = \frac{1}{p_T}\frac{dN}{dp_T} \simeq C\, e^{-m_T/T_{eff}}.$$ (5.214)

The $\langle m_T \rangle$ could be obtained by

$$\langle m_T \rangle = \frac{\int_0^\infty p_T\, dp_T\, m_T\, \exp(-m_T/T_{eff})}{\int_0^\infty p_T\, dp_T\, \exp(-m_T/T_{eff})}$$

$$= \frac{2T_{eff}^2 + 2m_0 T_{eff} + m_0^2}{m_0 + T_{eff}}$$ (5.215)

$$= T_{eff} + m_0 + \frac{(T_{eff})^2}{m_0 + T_{eff}}$$ (5.216)

$$\Rightarrow \boxed{\langle m_T \rangle = 2T_{eff} + \frac{m_0^2}{m_0 + T_{eff}}}$$ (5.217)

where m_0 is the rest mass of the particle. It can be seen from the above expression that for a massless particle

$$\langle m_T \rangle = \langle p_T \rangle = 2T_{eff}.$$ (5.218)

This also satisfies the principle of equipartition of energy, which is expected for a massless Boltzmann gas in equilibrium. However, in experiments, the lower (higher) limit of p_T is a finite quantity. In that case, the integration will involve an incomplete gamma function.

5.2.8 How Is the Radial Flow Measured from p_T-spectra?

In central heavy-ion collisions, radial flow is supposed to play a vital role in the thermodynamic expansion of the produced fireball. Radial flow is related to the initial pressure produced just after the collision. This could be extracted from the analysis of the transverse momentum spectra. Assuming a thermalized non-relativistic plasma (for simplicity), particle velocity [17]

$$\mathbf{v} = \mathbf{v}_{flow} + \mathbf{v}_{th},$$ (5.219)

where $\mathbf{v}_{flow} \equiv$ transverse velocity of the expanding fluid, which is independent of the particle species and is the collective component of $\mathbf{v}$.

$\mathbf{v}_{th} \equiv$ thermal component of $\mathbf{v}$, which is generated due to random thermal motion of the quanta of the system.

Hence, for a particle of rest mass m_0

$$\left\langle \frac{1}{2} m_0 v^2 \right\rangle = \frac{1}{2} m_0 v_{flow}^2 + \left\langle \frac{1}{2} m_0 v_{th}^2 \right\rangle$$

$$= \frac{1}{2} m_0 v_{flow}^2 + \frac{3}{2} k_B T, \tag{5.220}$$

where T is the temperature of the fluid and k_B is the Boltzmann constant. Hence, the average kinetic energy (K.E) is given by

$$\langle K.E. \rangle = \frac{1}{2} m_0 v_{flow}^2 + \frac{3}{2} k_B T$$

$$\Rightarrow \frac{3}{2} k_B T_{eff} = \frac{1}{2} m_0 v_{flow}^2 + \frac{3}{2} k_B T_{th}$$

$$\Rightarrow T_{eff} = T_{th} + \frac{1}{3} m_0 v_{flow}^2 \text{ (taking } k_B = 1) \tag{5.221}$$

Because the final-state particles are measured at freeze out (after they stream out to reach the detectors), the extracted values of $\mathbf{v}_{flow}$ and T correspond to the instant of freeze out.

Taking $k_B = 1$, in 2-dimension:

$$\boxed{T_{eff} = T_{th} + \frac{1}{2} m_0 v_{flow}^2}, \tag{5.222}$$

and in 1-dimension:

$$\boxed{T_{eff} = T_{th} + m_0 v_{flow}^2} \tag{5.223}$$

Note that these formulae are used essentially when the spectra are well described by an exponential function (low-p_T regime). However, when one goes to high-p_T regime, the following formula could be used for extracting the radial flow from the p_T or m_T-spectra.

$$\boxed{T_{eff} = T_{th} \sqrt{\frac{1 + v_{flow}}{1 - v_{flow}}}} \tag{5.224}$$

Lorentz Boost in Arbitrary Direction 

Consider S and S' to be two inertial frames, where S'-frame moves with a velocity $\mathbf{v}$ in an arbitrary direction unlike that we discussed earlier, which was along the X-axis, i.e., the velocity earlier was $\mathbf{v} = v\hat{\imath}$. At time $t = t' = 0$, both the origins and the axes of the Cartesian coordinate systems coincide (Fig. A.1).

Let $\mathbf{r}$ be the position vector of a point in the frame S, which is given by $\mathbf{r} = x\hat{\imath} + y\hat{\jmath} + z\hat{k}$. We decompose the spatial vector $\mathbf{r}$ into components perpendicular and parallel to $\mathbf{v}$ (Fig. A.2).

$$\mathbf{r} = \mathbf{r}_{||} + \mathbf{r}_{\perp} \tag{A.1}$$

so that

$$\mathbf{r}.\mathbf{v} = \mathbf{r}_{||} \cdot \mathbf{v} + \mathbf{r}_{\perp} \cdot \mathbf{v} = r_{||}v \tag{A.2}$$

Since $\mathbf{r}_{||}$ and $\mathbf{v}$ are parallel, we have

$$\mathbf{r}_{||} = r_{||}\frac{\mathbf{v}}{v} = \left(\frac{\mathbf{r}.\mathbf{v}}{v}\right)\frac{\mathbf{v}}{v}, \tag{A.3}$$

© The Author(s), under exclusive license to Springer Nature Switzerland AG 2025
R. Sahoo, *Relativistic Kinematics*, Lecture Notes in Physics 1046,
https://doi.org/10.1007/978-3-032-09510-7

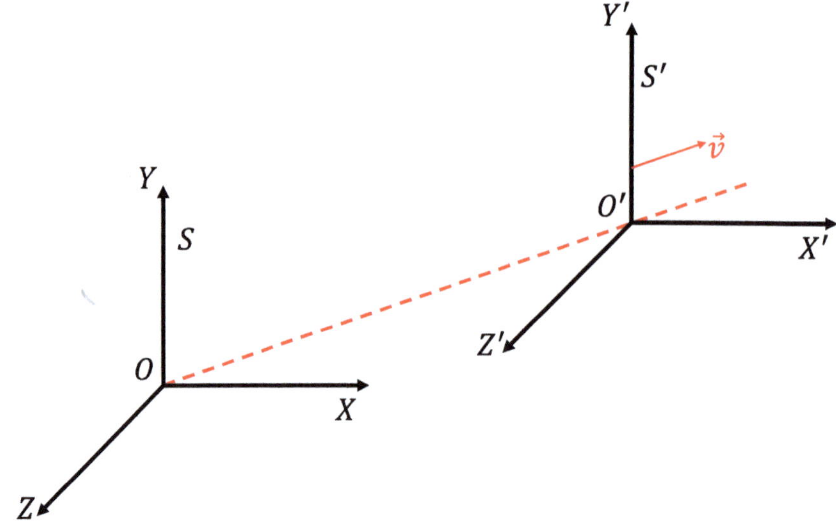

Fig. A.1 Lorentz boost in an arbitrary direction

Fig. A.2 Vector decomposition of the position vector along the boost direction and transverse to it

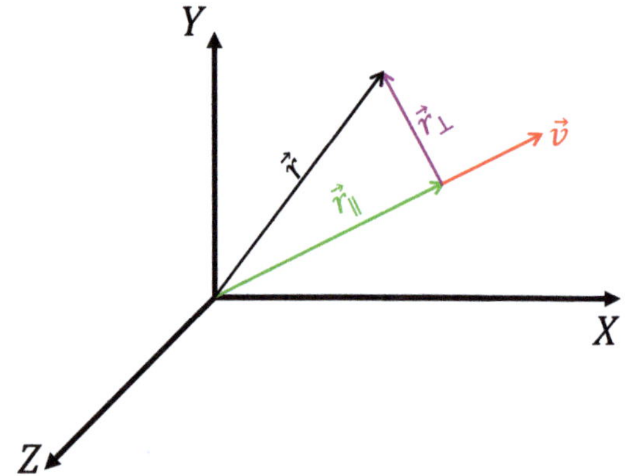

$\frac{\mathbf{v}}{v}$ is the unit vector in the direction of $\mathbf{r}_{\|}$ and $r_{\|} = \frac{\mathbf{r.v}}{v}$ is the projection of $\mathbf{r}$ in the direction of $\mathbf{v}$.

$$\mathbf{r}_{\|} = \frac{(\mathbf{r} \cdot \mathbf{v})}{|\mathbf{v}|^2}\mathbf{v}$$

$$= \frac{(\mathbf{r} \cdot \boldsymbol{\beta})}{\beta^2}\boldsymbol{\beta} \tag{A.4}$$

$$\mathbf{r}_\perp = \mathbf{r} - \frac{(\mathbf{r} \cdot \mathbf{v})}{|\mathbf{v}|^2}\mathbf{v}$$

$$= \mathbf{r} - \frac{(\mathbf{r} \cdot \boldsymbol{\beta})}{\beta^2}\boldsymbol{\beta} \tag{A.5}$$

The transverse component $\mathbf{r}_\perp$ is invariant under a Lorentz boost. However, the longitudinal component undergoes the usual transformation, with the boost factor:

$$\gamma = \frac{1}{\sqrt{1 - v^2/c^2}} = \frac{1}{\sqrt{1 - \beta^2}} \tag{A.6}$$

Now, the Lorentz transformation equations are:

$$t' = \gamma\left(t - \frac{\mathbf{r}.\mathbf{v}}{c^2}\right)$$

$$\Rightarrow ct' = \gamma(ct - \mathbf{r}.\boldsymbol{\beta}) \tag{A.7}$$

The transformation of the temporal coordinate is the same as a boost in the x-direction, but with $|\mathbf{r}||\mathbf{v}| = \mathbf{r} \cdot \mathbf{v}$ in place of vx.

$$\mathbf{r}' = \mathbf{r}_\perp + \gamma(\mathbf{r}_{||} - \mathbf{v}t) \tag{A.8}$$

$$\Rightarrow \mathbf{r}' = (\mathbf{r} - \mathbf{r}_{||}) + \gamma\mathbf{r}_{||} - \gamma\mathbf{v}t$$

$$\mathbf{r}' = \mathbf{r} + (\gamma - 1)\mathbf{r}_{||} - \gamma\boldsymbol{\beta}ct \tag{A.9}$$

It can also be written explicitly in terms of the position vector and boost velocity as:

$$\mathbf{r}' = \gamma\left(\frac{(\mathbf{r} \cdot \boldsymbol{\beta})}{\beta^2}\boldsymbol{\beta} - \boldsymbol{\beta}ct\right) + \left(\mathbf{r} - \frac{(\mathbf{r} \cdot \boldsymbol{\beta})}{\beta^2}\boldsymbol{\beta}\right) \tag{A.10}$$

In the matrix form:

$$\begin{bmatrix} ct' \\ \mathbf{r}' \end{bmatrix} = \begin{bmatrix} \gamma & -\gamma\boldsymbol{\beta}^T \\ -\gamma\boldsymbol{\beta} & \mathbf{I} + \frac{(\gamma-1)\boldsymbol{\beta}\boldsymbol{\beta}^T}{\beta^2} \end{bmatrix} \begin{bmatrix} ct \\ \mathbf{r} \end{bmatrix} \tag{A.11}$$

$\mathbf{I}$: 3×3 identity matrix and $\boldsymbol{\beta} = \frac{\mathbf{v}}{c}$ relative velocity vector, a column vector.

$$\boldsymbol{\beta} = \frac{\mathbf{v}}{c} = \begin{bmatrix} \beta_x \\ \beta_y \\ \beta_z \end{bmatrix} = \frac{1}{c}\begin{bmatrix} v_x \\ v_y \\ v_z \end{bmatrix} \equiv \begin{bmatrix} \beta_1 \\ \beta_2 \\ \beta_3 \end{bmatrix} = \frac{1}{c}\begin{bmatrix} v_1 \\ v_2 \\ v_3 \end{bmatrix}$$

$\boldsymbol{\beta}^T = \mathbf{v}/c$ is the transpose of $\boldsymbol{\beta}$, a row vector.

$$\boldsymbol{\beta}^T = \frac{\mathbf{v}^T}{c} = \begin{bmatrix} \beta_x & \beta_y & \beta_z \end{bmatrix} = \frac{1}{c}\begin{bmatrix} v_x & v_y & v_z \end{bmatrix} \equiv \begin{bmatrix} \beta_1 & \beta_2 & \beta_3 \end{bmatrix} = \frac{1}{c}\begin{bmatrix} v_1 & v_2 & v_3 \end{bmatrix}$$

β being the magnitude of $\boldsymbol{\beta}$,

$\beta = |\boldsymbol{\beta}| = [\beta_x^2 + \beta_y^2 + \beta_z^2]^{1/2}$

Explicitly:

$$\begin{bmatrix} ct' \\ x' \\ y' \\ z' \end{bmatrix} = \begin{bmatrix} \gamma & -\gamma\beta_x & -\gamma\beta_y & -\gamma\beta_z \\ -\gamma\beta_x & 1+(\gamma-1)\frac{\beta_x^2}{\beta^2} & (\gamma-1)\frac{\beta_x\beta_y}{\beta^2} & (\gamma-1)\frac{\beta_x\beta_z}{\beta^2} \\ -\gamma\beta_y & (\gamma-1)\frac{\beta_y\beta_x}{\beta^2} & 1+(\gamma-1)\frac{\beta_y^2}{\beta^2} & (\gamma-1)\frac{\beta_y\beta_z}{\beta^2} \\ -\gamma\beta_z & (\gamma-1)\frac{\beta_z\beta_x}{\beta^2} & (\gamma-1)\frac{\beta_z\beta_y}{\beta^2} & 1+(\gamma-1)\frac{\beta_z^2}{\beta^2} \end{bmatrix} \begin{bmatrix} ct \\ x \\ y \\ z \end{bmatrix}$$

$\Rightarrow \mathbf{X'} = \boldsymbol{\Lambda}(\mathbf{v})\mathbf{X}$

having structure:

$$\begin{bmatrix} ct \\ x' \\ y' \\ z' \end{bmatrix} = \begin{bmatrix} \Lambda_{00} & \Lambda_{01} & \Lambda_{02} & \Lambda_{03} \\ \Lambda_{10} & \Lambda_{11} & \Lambda_{12} & \Lambda_{13} \\ \Lambda_{20} & \Lambda_{21} & \Lambda_{22} & \Lambda_{23} \\ \Lambda_{30} & \Lambda_{31} & \Lambda_{32} & \Lambda_{33} \end{bmatrix} \begin{bmatrix} ct \\ x \\ y \\ z \end{bmatrix}$$

The components are:

$$\Lambda_{00} = \gamma$$

$$\Lambda_{0i} = \Lambda_{i0} = -\gamma\beta_i$$

$$\Lambda_{ij} = \Lambda_{ji} = (\gamma-1)\frac{\beta_i\beta_j}{\beta^2} + \delta_{ij} = (\gamma-1)\frac{v_iv_j}{v^2} + \delta_{ij}$$

where δ_{ij}: kronecker delta

$$\delta_{ij} = \begin{cases} 0 \ \ if \ \ i \neq j \\ 1 \ \ if \ \ i = j \end{cases}$$

Indeed, additionally, one can express the boost in an arbitrary direction as the product of rotations and the boost along the $X-$axis, which we don't discuss here.

References

1. D.H. Perkins, *Introduction to High Energy Physics*, 4th edn. (Cambridge University Press, Cambridge, 2000)
2. V.V. Skokov, A.Y. Illarionov, V.D. Toneev, Int. J. Mod. Phys. A **24**, 5925 (2009)
3. P. Romatschke, U. Romatschke, Phys. Rev. Lett. **99**, 172301 (2007)
4. https://cds.cern.ch/record/2800984?ln=en (2022). Accessed 27 Aug 2025
5. W.A. Barletta, et al., Accelerator capabilities, in *Planning the Future of U.S. Particle Physics: Proceedings of the Snowmass 2013 Community Study*, ed. by N.A. Graf, M.E. Peskin, J.L. Rosner (2014). https://www.slac.stanford.edu/econf/C1307292/ [arXiv:1401.6114]
6. R. Hagedorn, *Relativistic Kinematics* (W. A. Benjamin, Los Angeles, 1964)
7. J. Adam, et al., (ALICE Collaboration), Phys. Lett. B **772**, 567 (2017)
8. C. Alt, et al., (NA49 Collaboration), Phys. Rev. C **77**, 024903 (2008)
9. S.V. Afanasiev, et al., (NA49 Collaboration), Phys. Rev. C **66**, 054902 (2002)
10. P. Carruthers, M. Duong-van, Phys. Rev. D **8**, 859 (1973)
11. C.Y. Wong, Phys. Rev. C **78**, 054902 (2008)
12. H. Petersen, M. Bleicher, PoS CPOD2006, 025 (2006)
13. L.D. Landau, Izv. Akad. Nauk Ser. Fiz. **17**, 51 (1953)
14. S.Z. Belen'kji, L.D. Landau, Nuovo Cim. **3**(suppl. 1), 15 (1956)
15. P. Carruthers, Ann. N.Y. Acad. Sci. **229**, 91 (1974)
16. E.V. Shuryak, Yadernaya Fisika **16**, 395 (1972)
17. R.S. Bhalerao, *Lecture Notes in SERC School on Nuclear Physics* (IoP, Bhubaneswar, 2006)
18. B. Abelev, et al., (ALICE Collaboration), Phys. Rev. Lett. **109**, 252301 (2012)
19. S. Acharya, et al., (ALICE Collaboration), Eur. Phys. J. C **81**, 256 (2021)
20. http://www.agsrhichome.bnl.gov/RHIC/Runs/. Accessed 27 Aug 2025
21. R. Fernow, *Introduction to Experimental Particle Physics* (Cambridge University Press, Cambridge, 2001), and references therein
22. B.I. Abelev, et al., (STAR Collaboration), Phys. Rev. C **75**, 064901 (2007)
23. L. van Hove, Phys. Letts. B **118**, 138 (1982)

If you have any concerns about our products,
you can contact us on
ProductSafety@springernature.com

In case Publisher is established outside the EU,
the EU authorized representative is:
Springer Nature Customer Service Center GmbH
Europaplatz 3, 69115 Heidelberg, Germany

Printed by Libri Plureos GmbH
in Hamburg, Germany